Recent Advances in Scientific Computing and Matrix Analysis

Editors: Xiao-Qing Jin, Hai-Wei Sun & Seak-Weng Vong

International Press
www.intlpress.com

高等教育出版社
HIGHER EDUCATION PRESS

Recent Advances in Scientific Computing and Matrix Analysis
Edited by:
Xiao-Qing Jin, University of Macau
Hai-Wei Sun, University of Macau
Seak-Weng Vong, University of Macau

ISBN: 978-1-57146-202-2

Printed in the United States of America.

15 14 13 12 11 1 2 3 4 5 6 7 8 9

Dedicated to the 30th Anniversary of

University of Macau

(1981–2011)

Scientific Committee:

Prof. Zhong-ci Shi,	Chinese Academy of Sciences
Prof. Raymond H. Chan,	Chinese University of Hong Kong
Prof. Xiao-qing Jin,	University of Macau
Prof. Michael K. Ng,	Hong Kong Baptist University
Prof. Fu-rong Lin,	Shantou University
Dr. Hai-wei Sun,	University of Macau

Local Organizers:

(http://www.umac.mo)

Prof. Xiao-qing Jin (Chair),	University of Macau
Dr. Hai-wei Sun (Secretary),	University of Macau
Prof. Che-man Cheng,	University of Macau
Prof. Deng Ding,	University of Macau
Dr. Seak-weng Vong,	University of Macau
Dr. Siu-long Lei,	University of Macau

Preface

An international workshop on scientific computing and matrix analysis took place at University of Macau from 28th to 30th of December 2009. It was organized by the Department of Mathematics, University of Macau, for the 20th anniversary of the **T**eam **O**n **T**oeplitz **S**ystems (TOTS). The TOTS was born in 1989 in Hong Kong. In that year, Dr. Raymond H. Chan published three important papers on iterative Toeplitz solvers in SIAM journals and had his first graduate student Xiao-qing Jin. Raymond's iterative algorithm is based on the conjugate gradient method together with circulant matrices as preconditioners. He proved in his first paper in 1989 that the complexity of the algorithm is only $O(n \log n)$ where n is the size of the Toeplitz system. The techniques and ideas developed there have now become a standard tool in the trade where most papers adopt and follow. For his contributions in iterative Toeplitz solvers and their applications, Raymond was awarded the Leslie Fox Prize in 1989, the Feng Kang Prize in Scientific Computing in 1997, and the Morningside Silver Medal of Mathematics in 1998.

Since 1989, iterative Toeplitz solvers have become a new research area in numerical linear algebra. With twenty years passed, the TOTS has produced several academic generations and obtained great academic achievements. The main theme of this workshop is the latest development in *matrix computations and applications* made by the TOTS and the papers involved in this proceedings cover this theme. There are sixteen invited speakers from USA, Canada, Singapore, Hong Kong, Macao, and mainland China. Twelve of them are the TOTS members. The following is the list of invited speakers:

Prof. Zhong-ci Shi of Chinese Academy of Sciences, Beijing, China.

Prof. Raymond H. Chan of Chinese University of Hong Kong, Hong Kong.

Prof. Zheng-jian Bai of Xiamen University, Xiamen, China.

Prof. Wai-ki Ching of University of Hong Kong, Hong Kong.

Dr. Yu-mei Huang of Lanzhou University, Lanzhou, China.

Dr. Li-ping Jing of Beijing Jiaotong University, Beijing, China.

Prof. Wen Li of South China Normal University, Guangzhou, China.

Dr. Hai-yong Liao of Hong Kong Baptist University, Hong Kong.

Prof. Fu-rong Lin of Shantou University, Shantou, China.

Prof. San-zheng Qiao of McMaster University, Canada.

Dr. Chi-pan Tam of Hong Kong Baptist University, Hong Kong.

Prof. Yi-min Wei of Fudan University, Shanghai, China.

Dr. You-wei Wen of South China Agricultural University, Guangzhou, China.

Prof. Man-chung Yeung of University of Wyoming, USA.

Dr. Andy M. Yip of National University of Singapore, Singapore.

Dr. Shu-qin Zhang of Fudan University, Shanghai, China.

There are eleven chapters (papers) in this proceedings. The main contents of chapters are described briefly as follows.

A Ulm-like method is proposed for solving inverse eigenvalue problems with distinct given eigenvalues in Chapter 1. The Ulm-like method avoids solving the Jacobian equations used in Newton-like methods and is shown to be quadratically convergent in the root sense. **Z. J. Bai** and X. Q. Jin give a numerical example to show that the Ulm-like method is better than the inexact Newton-like method in terms of convergence neighborhoods.

It has been proved that for any $n \times n$ complex matrices X and Y,

$$\|XY - YX\|_F \leq \sqrt{2}\|X\|_F\|Y\|_F, \qquad (*)$$

where $\|\cdot\|_F$ denotes the Frobenius norm. A characterization of those pairs of matrices that satisfy $(*)$ has also been found. Recently, K. Audenaert has given a new proof of the inequality by introducing a matrix version of variance. In Chapter 2, based on his proof, C. M. Cheng, K. S. Fong, and I. K. Luk give another proof of the equality of $(*)$.

W. K. Ching and D. M. Zhu give a brief review in Chapter 3 on some recent models and results in high-dimensional Markov chain models for categorical data sequences. Categorical data sequences occur in many areas including bioinformatics, operations research, data mining, and financial data modeling, etc. The construction of mathematical models for categorical data sequences is important in the sense that it can help in the discovery of rules, testing hypothesis, and making effective predictions. Ching and Zhu introduce a number of high-dimensional Markov chain models with applications. Efficient estimation methods for the model parameters are also discussed.

In Chapter 4, Y. N. Law, H. K. Lee, C. Q. Liu, and **A. M. Yip** propose a variant of the Mumford-Shah model for the segmentation of overlapping semi-transparent objects with additive intensity value. Unlike standard segmentation models, it does not only determine distinct objects in the image, but also recover possibly multiple membership of the pixels. To accomplish this, *a priori* knowledge about the smoothness of the objects is taken into account in the model. To solve the optimization problem involving geometric quantities efficiently, they apply a multi-phase level set method. Segmentation results on synthetic and real images validate the good performance of their model.

Total variation based methods are popular for image restoration. One typical problem in this kind of method is how to select the regularization parameters. The common approach is try-and-error. In the literature there are several regularization parameter selection methods for Tikhonov regularization problems. In Chapter 5, **H. Y. Liao** and M. K. Ng study the regularization parameter problems in total variation based image deblurring problem. Within the alternating minimization framework, they exploit the generalized cross-validation technique to select the regularization parameter in each iteration. Liao and Ng also extend the idea to blind deconvolution problem. The experimental results are given to demonstrate the performance of the proposed methods.

F. T. Luk and **S. Z. Qiao** present a matrix interpretation of the LLL algorithm for lattice basis reduction in Chapter 6.

A nonconvex nonsmooth regularization has advantages over a convex regularization for restoring images with neat edges. However, its practical interest used to be limited by the difficulty of the computational stage which requires a nonconvex nonsmooth minimization. In Chapter 7, M. Nikolova, M. K. Ng, and **C. P. Tam** study a fast nonconvex nonsmooth minimization method for image restoration and reconstruction. The main aim is to develop a fast minimization algorithm to solve the nonconvex nonsmooth minimization problem. The experimental results show that the effectiveness and efficiency of the proposed method.

Given a row stochastic matrix P of size n, G. Wu examines the relationship between the eigenvalues of P and those of some augmented matrices with respect to P in Chapter 8. Applications of the results to Google's PageRank problem are also discussed.

Chapter 9 considers numerical solution methods for Wiener-Hopf equations of the second kind defined on $[0, \infty)$:

$$y(t) + \int_0^\infty k(t-s)y(s)\mathrm{d}s = g(t), \qquad 0 \le t < \infty.$$

Y. Xuan and **F. R. Lin** first combine the well-known Clenshaw-Curtis quadrature with a rational variable transformation to get a Clenshaw-Curtis-rational (CCR) quadrature rule, which is suitable for integrals on semi-infinite intervals. They then apply the CCR quadrature to Wiener-Hopf equations. The reduction of singularities in Wiener-Hopf equations is also discussed. Numerical examples are given to illustrate the efficiency of the CCR quadrature.

Krylov subspace methods are popular iterative methods to solve large sparse linear systems in the real-world computations due to their cheap memory requirement and computational cost. In Chapter 10, **M. C. Yeung** discusses the solution of singular systems. He shows that the consistency of a singular linear system is not a sufficient condition for a Krylov subspace method to successfully find a solution of the system. The choice of initial guess is a crucial step. If the initial guess is properly chosen, a Krylov method almost surely converges to find a solution from the viewpoint of probability, otherwise a Krylov subspace method surely diverges.

For solving the integer least squares problems arising in many engineering applications, the LLL algorithm is widely used as a preconditioner. But the way how the LLL algorithm works was seldom understood by practitioners, until Luk and Tracy first described the behavior of the algorithm in 2008, and they also derived a new numerical implementation of the LLL algorithm. Luk and Qiao then compare the original and the new LLL algorithms in the cases of overflow and underflow. In Chapter 11, Q. F. Yu, **S. Z. Qiao**, and **Y. M. Wei** look into the difference when the LLL algorithms are applied to the integer least squares problems.

The success of the workshop should be credited to the enthusiastic participation of the attendees. Special thanks for a fund to workshop are addressed to University of Macau. The hospitality of the local organizers and helpers is greatly appreciated. The editors would like to thank all who contributed to this proceedings of the workshop. We wish that the proceedings may provide some inspiration and shed new light on recent research in matrix computations and applications. The publication of the proceedings is supported by the research grant UL020/08-Y3/MAT/JXQ01/FST from University of Macau.

CONTENTS

Chapter 1

A Note on the Ulm-like Method for Inverse Eigenvalue Problems

Zheng-jian Bai [1] **Xiao-qing Jin** [2]

Abstract

A Ulm-like method is proposed in [12] for solving inverse eigenvalue problems with distinct given eigenvalues. The Ulm-like method avoids solving the Jacobian equations used in Newton-like methods and is shown to be quadratically convergent in the root sense. However, the numerical experiments in [12] only show that the Ulm-like method is comparable with the inexact Newton-like method in [4]. In this chapter, we give a numerical example to show that the Ulm-like method is better than the inexact Newton-like method in terms of convergence neighborhoods.

1.1 Introduction

Let $A_1, A_2, \ldots, A_n$ be n real symmetric n-by-n matrices. For any $\mathbf{c} = (c_1, \ldots, c_n)^T \in \mathbb{R}^n$, let $\{\lambda_i(\mathbf{c})\}_{i=1}^n$ denote the eigenvalues of the matrix

$$A(\mathbf{c}) \equiv \sum_{i=1}^{n} c_i A_i,$$

where $\lambda_1(\mathbf{c}) \le \lambda_2(\mathbf{c}) \le \cdots \le \lambda_n(\mathbf{c})$. In this chapter, we consider the inverse eigenvalue problem (IEP) defined as follows:

IEP: Given n real numbers $\lambda_1^* \le \cdots \le \lambda_n^*$, find a vector $\mathbf{c}^* \in \mathbb{R}^n$ such that $\lambda_i(\mathbf{c}^*) = \lambda_i^*$ for $i = 1, \ldots, n$.

---

[1] School of Mathematical Sciences, Xiamen University, Xiamen 361005, P. R. China (email: zjbai@xmu.edu.cn). The research of this author is partially supported by the Natural Science Foundation of Fujian Province of China for Distinguished Young Scholars, NCETXMU, and SRF for ROCS, SEM.

[2] Department of Mathematics, University of Macau, Macao (email: xqjin@umac.mo). The research of this author is supported by the grant UL020/08-Y3/MAT/JXQ01/FST from University of Macau.

The inverse eigenvalue problem arises in many applications, such as the solution of inverse Sturm-Liouville problems, nuclear and molecular spectroscopy, applied mechanics and structure design, etc. For applications, theoretical and algorithmic aspects of general inverse eigenvalue problems, see [5, 6, 7, 9, 10, 13].

Assume that the given eigenvalues are distinct and there exists a solution $\mathbf{c}^*$ for the IEP. Then there exists a neighborhood $\mathcal{N}(\mathbf{c}^*)$ of $\mathbf{c}^*$ such that $\lambda_i(\mathbf{c})$ are distinct and differentiable [13, Theorem 1.9.2]. In this neighborhood $\mathcal{N}(\mathbf{c}^*)$, the IEP can be written as a nonlinear equation

$$\mathbf{f}(\mathbf{c}) \equiv (\lambda_1(\mathbf{c}) - \lambda_1^*, \ldots, \lambda_n(\mathbf{c}) - \lambda_n^*)^T = \mathbf{0}, \quad \forall \mathbf{c} \in \mathcal{N}(\mathbf{c}^*). \tag{1.1}$$

It is well-known [9, 13] that Newton-like methods can be employed for solving the IEP based on (1.1). For inexact Newton-like approaches, see [1, 2, 3, 4] for instance. In each outer Newton iteration, however, we need to solve the (approximate) Jacobian equation, which may cause the instability problem if the Jacobian matrix is ill-conditioned.

In [12], a Ulm-like method is proposed for the IEP. The advantage of the Ulm-like method is replacing the solution of the (approximate) Jacobian equations by computing the product of matrices. Therefore, the Ulm-like method avoids the instability problem caused by the possible ill-conditioning in solving the (approximate) Jacobian equations. The Ulm-like method has a root-quadratic convergence rate under the assumption that the given eigenvalues are distinct.

However, the numerical tests in [12] only show that the Ulm-like method is comparable with the inexact Newton-like method in [4] in terms of outer Newton iterations. In this chapter, we present an example to show that the Ulm-like method is better than the inexact Newton-like method in terms of convergence neighborhoods.

1.2 Inexact Newton-like method and Ulm-like method

We first recall the inexact Newton-like method and the Ulm-like method, respectively, for solving the IEP. Let $\|\cdot\|$ denote the Euclidean norm in $\mathbb{R}^n$ and the corresponding induced matrix norm in $\mathbb{R}^{n\times n}$, i.e.,

$$\|A\| \equiv \max\left\{\frac{\|A\mathbf{x}\|}{\|\mathbf{x}\|} : \mathbf{x} \in \mathbb{R}^n \text{ with } \mathbf{x} \neq \mathbf{0}\right\}.$$

Suppose that $\mathbf{c}^*$ is the solution of the IEP. For any $\mathbf{c} \in \mathbb{R}^n$, let $\mathbf{p}_k(\mathbf{c})$ be the normalized eigenvector of $A(\mathbf{c})$ corresponding to $\lambda_k(\mathbf{c})$ for $k = 1, 2, .., n$. The matrix $J(\mathbf{c})$ is defined by

$$[J(\mathbf{c})]_{ij} \equiv \mathbf{p}_i(\mathbf{c})^T A_j \mathbf{p}_i(\mathbf{c}),$$

for $i, j = 1, 2, \ldots, n$. Let

$$\boldsymbol{\lambda}^* = (\lambda_1^*, \lambda_2^*, \ldots, \lambda_n^*)^T,$$

where $\lambda_1^* \leq \lambda_2^* \leq \cdots \leq \lambda_n^*$.

The inexact Newton-like algorithm [4] is given as follows.

Algorithm I: Inexact Newton-like Method

1. *Given* $\mathbf{c}^0$, *compute the orthonormal eigenvectors* $\{\mathbf{p}_i^0 = \mathbf{p}_i(\mathbf{c}^0)\}_{i=1}^n$ *of* $A(\mathbf{c}^0)$ *and solve* $\mathbf{c}^1$ *by*

$$J(\mathbf{c}^0)\mathbf{c}^1 = \boldsymbol{\lambda}^*.$$

2. *For* $k = 0, 1, 2, \ldots,$ *until convergence, do:*

   (a) *Solve* $\mathbf{v}_i^k$ *inexactly by the one-step inverse power method:*

$$\big(A(\mathbf{c}^k) - \lambda_i^* I\big)\mathbf{v}_i^k = \mathbf{p}_i^{k-1} + \mathbf{t}_i^k, \quad i = 1, 2, \ldots, n, \tag{1.2}$$

   where I *is the* n*-by-*n *identity matrix, until the residual* $\mathbf{t}_i^k$ *satisfies*

$$\|\mathbf{t}_i^k\| \le \frac{1}{4}, \quad i = 1, 2, \ldots, n.$$

   (b) *Normalize* $\mathbf{v}_i^k$ *to obtain an approximate eigenvector* $\mathbf{p}_i^k$ *of* $A(\mathbf{c}^k)$*:*

$$\mathbf{p}_i^k = \frac{\mathbf{v}_i^k}{\|\mathbf{v}_i^k\|}, \quad i = 1, 2, \ldots, n.$$

   (c) *Form the approximate Jacobian matrix* J_k*:*

$$[J_k]_{ij} = (\mathbf{p}_i^k)^T A_j \mathbf{p}_i^k, \quad i, j = 1, 2, \ldots, n. \tag{1.3}$$

   (d) *Solve* $\mathbf{c}^{k+1}$ *inexactly from the approximate Jacobian equation:*

$$J_k \mathbf{c}^{k+1} = \boldsymbol{\lambda}^* + \mathbf{r}^k \tag{1.4}$$

   until the residual $\mathbf{r}^k$ *satisfies*

$$\|\mathbf{r}^k\| \le \left(\max_{1\le i\le n} \frac{1}{\|\mathbf{v}_i^k\|}\right)^\beta, \quad \text{for} \quad 1 < \beta \le 2.$$

Under the assumptions that the given eigenvalues are distinct and the Jacobian matrix $J(\mathbf{c}^*)$ is nonsingular, it was proved in [4] that the method converges locally with a convergence rate β.

Next, we present the Ulm-like method for the IEP.

Algorithm II: Ulm-like Method

1. *Let* $\mathbf{c}^0 \in \mathbb{R}^n$ *and* $B_0 \in \mathbb{R}^{n\times n}$ *be given such that*

$$\|I - B_0 J(\mathbf{c}^0)\| \le \delta,$$

   where δ *is a positive constant. Compute the orthonormal eigenvectors* $\{\mathbf{p}_i^0 = \mathbf{p}_i(\mathbf{c}^0)\}_{i=1}^n$ *of* $A(\mathbf{c}^0)$ *and solve* $\mathbf{c}^1$ *by*

$$\mathbf{c}^1 = \mathbf{c}^0 - B_0\big(J(\mathbf{c}^0)\mathbf{c}^0 - \boldsymbol{\lambda}^*\big).$$

2. *For* $k = 0, 1, 2, \ldots,$ *until convergence, do:*

   (a) *Solve* $\mathbf{v}_i^k$ *inexactly by the one-step inverse power method:*

$$\big(A(\mathbf{c}^k) - \lambda_i^* I\big)\mathbf{v}_i^k = \mathbf{p}_i^{k-1} + \mathbf{t}_i^k, \quad i = 1, 2, \ldots, n, \tag{1.5}$$

   until the residual $\mathbf{t}_i^k$ *satisfies* $\|\mathbf{t}_i^k\| \le 1/4$ *for* $i = 1, 2, \ldots, n$.

(b) *Normalize* $\mathbf{v}_i^k$ *to obtain an approximate eigenvector* $\mathbf{p}_i^k$ *of* $A(\mathbf{c}^k)$:

$$\mathbf{p}_i^k = \frac{\mathbf{v}_i^k}{\|\mathbf{v}_i^k\|}, \quad i = 1, 2, \ldots, n.$$

(c) *Form the approximate Jacobian matrix* J_k:

$$[J_k]_{ij} = (\mathbf{p}_i^k)^T A_j \mathbf{p}_i^k, \quad i, j = 1, 2, \ldots, n.$$

(d) *Compute the matrix* B_k *and* $\mathbf{c}^{k+1}$ *respectively by*

$$B_k = 2B_{k-1} - B_{k-1} J_k B_{k-1}$$

and

$$\mathbf{c}^{k+1} = \mathbf{c}^k - B_k(J_k \mathbf{c}^k - \boldsymbol{\lambda}^*).$$

Under the assumptions that the given eigenvalues are distinct and the Jacobian matrix $J(\mathbf{c}^*)$ is nonsingular, the Ulm-like method converges quadratically in the root sense [12].

1.3 Numerical tests

In this section, we compare the numerical performance of Algorithm I with that of Algorithm II on an example below. Our goal is to illustrate the advantage of the Ulm-like method over the inexact Newton-like method, when the approximate Jacobian equation (1.4) is ill-conditioned.

Since the given eigenvalues are assumed to be distinct, the function $\mathbf{f}(\mathbf{c})$ in (1.1) is analytic in the neighborhood $\mathcal{N}(\mathbf{c}^*)$ of $\mathbf{c}^*$ and the Jacobian matrix of $\mathbf{f}$ is given by ([13, Theorem 1.9.2])

$$J(\mathbf{c}) = \mathbf{f}'(\mathbf{c}) = [\mathbf{p}_i(\mathbf{c})^T A_j \mathbf{p}_i(\mathbf{c})].$$

For $\mathbf{c} \in \mathcal{N}(\mathbf{c}^*)$, by (see [11])

$$\frac{\partial^2 \lambda_i}{\partial c_k \, \partial c_j} = 2 \sum_{\substack{m=1 \\ m \neq i}}^{n} \frac{\left(\mathbf{p}_m(\mathbf{c})^T A_k \mathbf{p}_i(\mathbf{c})\right)\left(\mathbf{p}_m(\mathbf{c})^T A_j \mathbf{p}_i(\mathbf{c})\right)}{\lambda_i(\mathbf{c}) - \lambda_m(\mathbf{c})},$$

it follows that $J(\mathbf{c})$ is Lipschitz continuous in the neighborhood $\mathcal{N}(\mathbf{c}^*)$ of $\mathbf{c}^*$, where the Lipschitz constant may be very large if some of the eigenvalues λ_i^* are close to each other. Therefore, for the inexact Newton-like method, the approximate Jacobian matrix J_k defined in (1.3) may be ill-conditioned and the convergence neighborhood $\mathcal{N}(\mathbf{c}^*)$ may be very small [9].

Example 1.1 *Given* $B = I_8 + VV^T$, *where* I_8 *is the* 8*-by-*8 *identity matrix and*

$$V = \begin{bmatrix} 1 & -1 & -3 & -5 & -6 \\ 1 & 1 & -2 & -5 & -17 \\ 1 & -1 & -1 & 5 & 18 \\ 1 & 1 & 1 & 2 & 0 \\ 1 & -1 & 2 & 0 & 1 \\ 1 & 1 & 3 & 0 & -1 \\ 2.5 & .2 & .3 & .5 & .6 \\ 2 & -.2 & .3 & .5 & .8 \end{bmatrix}_{8\times 5}$$

Define A_k by

$$A_k = b_{kk}\mathbf{e}_k\mathbf{e}_k^T + \sum_{j=1}^{k-1} b_{kj}(\mathbf{e}_k\mathbf{e}_j^T + \mathbf{e}_j\mathbf{e}_k^T), \quad k = 1, 2, \ldots, 8,$$

where $\mathbf{e}_k$ is the kth *column of I_8. Note that if $\hat{\mathbf{c}} = (1,1,1,1,1,1,1,1)^T$, then*

$$A(\hat{\mathbf{c}}) = B = I_8 + VV^T,$$

whose eigenvalues are given by

$$\hat{\boldsymbol{\lambda}} = (1, 1, 1, 2.1208, 9.2189, 17.2814, 35.7082, 722.6808)^T.$$

Now, suppose that

$$\begin{aligned}\mathbf{c}^* &= (1.000438903816714, 1.000656447518457, 1.000913442705718, 1.000231554995865,\\ &\quad 0.999744815493349, 0.999113996722789, 1.000942919907134, 0.999654879193127)^T.\end{aligned}$$

Then, we have

$$\begin{aligned}\boldsymbol{\lambda}^* &= (0.9793644297787, 0.9976265969314, 1.0039322015831, 2.1258971800068,\\ &\quad 9.2125235810642, 17.2782020459764, 35.6897669639946, 723.2816411319387)^T.\end{aligned}$$

In Example 1.1, we observe that the first three eigenvalues, λ_1, λ_2, and λ_3, are distinct but slightly close to each other:

$$|\lambda_2 - \lambda_1| = 1.8262e-02, \qquad |\lambda_3 - \lambda_2| = 6.3056e-03.$$

Thus, the approximate Jacobian matrix J_k in (1.3) may be ill-conditioned and the convergence neighborhood $\mathcal{N}(\mathbf{c}^*)$ becomes small.

To illustrate the numerical performance of the Ulm-like method, we compare the Ulm-like method with the inexact Newton-like method for Example 1.1. The initial point $\mathbf{c}^0$ is chosen by

$$\mathbf{c}^0 = \frac{\texttt{floor}(\phi \times 10^{\psi} \cdot \mathbf{c}^*)}{\phi \times 10^{\psi}},$$

where the `Matlab`-routine function `floor(c)` rounds the elements of a vector $\mathbf{c}$ to the nearest integers towards $-\infty$. We consider the following four cases: (a) $\phi = 3$ and $\psi = 1$; (b) $\phi = 4$ and $\psi = 1$; (c) $\phi = 1$ and $\psi = 2$; (d) $\phi = 1$ and $\psi = 4$.

The linear systems (1.2), (1.4), and (1.5) are solved iteratively by the QMR method [8] as in [1, 4, 12] using the `Matlab`-provided `QMR` function. At the $(k+1)$th iteration, we use $\mathbf{v}_i^k$ as the initial guess of the inverse power equations (1.2) and (1.5), and $\mathbf{c}^k$ as the initial guess of the approximate Jacobian equation (1.4). We also set the maximum number of iterations allowed to be 400 for all inner iterations. To speed up the convergence, we use the `Matlab`-provided Modified ILU (MILU) preconditioner: `LUINC`(A, [drop-tolerance, 1, 1, 1]), which is used for unconstructed matrices [1, 3, 12]. The drop-tolerance we use here is 0.01. The stopping tolerance for the outer Newton iterations of Algorithms I and II is set to be 10^{-10}, and the maximum number of outer iterations is set to be 10.

Table 1.1 displays the error of $\|\mathbf{c}^k - \mathbf{c}^*\|$ and the condition number $\kappa_2(J_k)$ for different initial guess $\mathbf{c}^0$, where $\mathbf{c}^k$, `it.`, and "*" denote the iterant at the kth outer iteration, the number of required outer iterations, and the corresponding algorithm fails to converge, respectively. We observe from Table 1.1 the following facts: When the initial guess $\mathbf{c}^0$ is somewhat far from the solution $\mathbf{c}^*$, the approximate Jacobian equation (1.4) is increasingly ill-conditioned and the inexact Newton-like method fails to converge, while the Ulm-like method still converges. On the other hand, when the initial guess $\mathbf{c}^0$ is close to $\mathbf{c}^*$ sufficiently, the approximate Jacobian equation (1.4) is well-conditioned. In this case, the Ulm-like method is comparable with the inexact Newton-like method. This numerical example shows that the Ulm-like method has a larger convergence neighborhood than that of the inexact Newton-like method.

References for Chapter 1

[1] Z.-J. Bai, R. H. Chan, and B. Morini, *An inexact Cayley transform method for inverse eigenvalue problems*, Inverse Problems, 20 (2004), pp. 1675–1689.

[2] Z.-J. Bai, *Inexact Newton methods for inverse eigenvalue problems*, Appl. Math. Comput., 172 (2006), pp. 682–689.

[3] Z.-J. Bai and S.-F. Xu, *An inexact Newton-type method for inverse singular value problems*, Linear Algebra Appl., 429 (2008), pp. 527–547.

[4] R. H. Chan, H.-L. Chung, and S.-F. Xu, *The inexact Newton-like method for inverse eigenvalue problem*, BIT, 43 (2003), pp. 7–20.

[5] M. T. Chu, *Inverse eigenvalue problems*, SIAM Rev., 40 (1998), pp. 1–39.

[6] M. T. Chu and G. H. Golub, *Structured inverse eigenvalue problems*, Acta Numer., 11 (2002), pp. 1–71.

[7] M. T. Chu and G. H. Golub, *Inverse Eigenvalue Problems: Theory, Algorithms, and Applications*, Oxford University Press, Oxford, 2005.

[8] R. W. Freund and N. M. Nachtigal, *QMR: A quasi-minimal residual method for non-Hermitian linear systems*, Numer. Math., 60 (1991), pp. 315–339.

[9] S. Friedland, J. Nocedal, and M. L. Overton, *The formulation and analysis of numerical methods for inverse eigenvalue problems*, SIAM J. Numer. Anal., 24 (1987), pp. 634–667.

[10] G. M. L. Gladwell, *Inverse Problems in Vibration*, Kluwer Academic Publishers, Dordrecht, 2004.

[11] P. Lancaster, *On eigenvalues of matrices dependent on a parameter*, Numer. Math., 6 (1964), pp. 377–387.

[12] W.-P. Shen, C. Li, and X.-Q. Jin, *A Ulm-like method for inverse eigenvalue problems*, Appl. Numer. Math., to appear.

[13] S.-F. Xu, *An Introduction to Inverse Algebraic Eigenvalue Problems*, Peking University Press and Vieweg Publishing, Beijing, 1998.

Table 1.1: Numerical results for Example 1.1.

(a)

k	Inexact Newton-like Method				Ulm-like Method	
	$\|\mathbf{c}^k - \mathbf{c}^*\|$					
	$\beta = 1.6$	$\beta = 1.8$	$\beta = 2.0$	$\kappa_2(J_k)$	$\|\mathbf{c}^k - \mathbf{c}^*\|$	$\kappa_2(J_k)$
0	$5.6900e-02$	$5.6900e-02$	$5.6900e-02$	$6.2633e+02$	$5.6900e-02$	$6.2633e+02$
1	$1.5972e-02$	$1.5972e-02$	$1.5972e-02$	$2.0217e+03$	$1.5972e-02$	$2.0217e+03$
2	$7.5108e-03$	$7.5108e-03$	$7.5108e-03$	$1.2617e+03$	$2.4305e-03$	$1.1030e+03$
3	$5.2259e-03$	$5.2259e-03$	$5.2259e-03$	$4.8507e+04$	$8.5988e-05$	$1.0936e+03$
4	$8.5108e-02$	$8.5108e-02$	$8.5108e-02$	$1.4192e+03$	$3.7376e-06$	$1.0850e+03$
5	$7.9658e-02$	$7.9658e-02$	$7.9658e-02$	$3.1942e+05$	$3.2939e-09$	$1.0852e+03$
6	$2.7836e+00$	$2.7836e+00$	$2.7836e+00$	$7.4580e+04$	$2.9768e-14$	
7	$2.7836e+00$	$2.7836e+00$	$2.7836e+00$	$5.5743e+06$		
8	$2.7836e+00$	$2.7836e+00$	$2.7836e+00$	$2.1713e+08$		
9	$2.7836e+00$	$2.7836e+00$	$2.7836e+00$	$1.2112e+09$		
10	$2.7836e+00$	$2.7836e+00$	$2.7836e+00$			
it.	*	*	*		6	

(b)

k	Inexact Newton-like Method				Ulm-like Method	
	$\|\mathbf{c}^k - \mathbf{c}^*\|$					
	$\beta = 1.6$	$\beta = 1.8$	$\beta = 2.0$	$\kappa_2(J_k)$	$\|\mathbf{c}^k - \mathbf{c}^*\|$	$\kappa_2(J_k)$
0	$4.2474e-02$	$4.2474e-02$	$4.2474e-02$	$6.3724e+02$	$4.2474e-02$	$6.3724e+02$
1	$1.5893e-02$	$1.5893e-02$	$1.5893e-02$	$1.8435e+03$	$1.5893e-02$	$1.8435e+03$
2	$5.8073e-03$	$5.8073e-03$	$5.8613e-03$	$1.1812e+03$	$2.3494e-03$	$1.1145e+03$
3	$2.7831e-03$	$2.7184e-03$	$2.1950e-03$	$1.4335e+03$	$2.7159e-04$	$1.1052e+03$
4	$2.1579e-04$	$1.8335e-04$	$8.7238e-05$	$1.0798e+03$	$5.7055e-06$	$1.0849e+03$
5	$2.5776e-06$	$1.6465e-06$	$4.2657e-07$	$1.0851e+03$	$7.6267e-09$	$1.0852e+03$
6	$4.6985e-10$	$1.3123e-10$	$8.7389e-12$		$1.6943e-14$	
7	$2.5663e-14$	$1.9658e-14$				
it.	7	7	6		6	

(c)

k	Inexact Newton-like Method				Ulm-like Method	
	$\|\mathbf{c}^k - \mathbf{c}^*\|$					
	$\beta = 1.6$	$\beta = 1.8$	$\beta = 2.0$	$\kappa_2(J_k)$	$\|\mathbf{c}^k - \mathbf{c}^*\|$	$\kappa_2(J_k)$
0	$1.6542e-02$	$1.6542e-02$	$1.6542e-02$	$6.6323e+02$	$1.6542e-02$	$6.6323e+02$
1	$1.6320e-02$	$1.6320e-02$	$1.6320e-02$	$1.7985e+03$	$1.6320e-02$	$1.7985e+03$
2	$5.2114e-03$	$5.2114e-03$	$5.3287e-03$	$1.1766e+03$	$1.3189e-03$	$1.0968e+03$
3	$1.7118e-03$	$1.6561e-03$	$1.4067e-03$	$1.2546e+03$	$2.0367e-04$	$1.0961e+03$
4	$6.2036e-05$	$5.2056e-05$	$3.7954e-05$	$1.0829e+03$	$1.2452e-05$	$1.0857e+03$
5	$2.4555e-07$	$1.4633e-07$	$7.7295e-08$	$1.0852e+03$	$2.9611e-08$	$1.0852e+03$
6	$2.0155e-11$	$9.9290e-13$	$2.4267e-13$		$1.0741e-13$	
it.	6	6	6		6	

(d)

k	Inexact Newton-like Method				Ulm-like Method	
	$\|\mathbf{c}^k - \mathbf{c}^*\|$					
	$\beta = 1.6$	$\beta = 1.8$	$\beta = 2.0$	$\kappa_2(J_k)$	$\|\mathbf{c}^k - \mathbf{c}^*\|$	$\kappa_2(J_k)$
0	$1.1373e-04$	$1.1373e-04$	$1.1373e-04$	$1.0856e+03$	$1.1373e-04$	$1.0856e+03$
1	$2.0352e-07$	$2.0352e-07$	$2.0352e-07$	$1.0852e+03$	$2.0352e-07$	$1.0852e+03$
2	$2.9669e-14$	$2.9669e-14$	$2.9669e-14$		$1.5068e-13$	
it.	2	2	2		2	

Chapter 2

Another Proof for Commutators with Maximal Frobenius Norm

Che-man Cheng [1] **Kin-sio Fong** [2] **Io-kei Lok** [3]

Abstract

It has been proved that for all $n \times n$ complex matrices X and Y,

$$\|XY - YX\|_F \leq \sqrt{2}\|X\|_F\|Y\|_F,$$

where $\|\cdot\|_F$ denotes the Frobenius norm. A characterization of those pairs of matrices that satisfy the corresponding equality

$$\|XY - YX\|_F = \sqrt{2}\|X\|_F\|Y\|_F$$

has also been found. Recently, Audenaert has given a new proof of the inequality by introducing a matrix version of variance. In this chapter, based on his proof, we give another proof for the characterization of the matrices that attain the above equality.

2.1 Introduction

In [2], Böttcher and Wenzel conjectured that for all $n \times n$ real matrices X and Y,

$$\|XY - YX\|_F \leq \sqrt{2}\|X\|_F\|Y\|_F, \tag{2.1}$$

where $\|\cdot\|_F$ denotes the Frobenius norm. It is shown to be true for 3×3 matrices by László [5], and for general $n \times n$ matrices by Vong and Jin [7] (independently by Lu [6]). The result has been extended to complex matrices by Böttcher and Wenzel [3]. In [4], Cheng et al. gave a complete characterization of the pairs of nonzero matrices which satisfy (2.1)

---

[1]Department of Mathematics, University of Macau, Macao, China (email: fstcmc@umac.mo).
[2]Department of Mathematics, University of Macau, Macao, China.
[3]Department of Mathematics, University of Macau, Macao, China.

with equality. Right after they found the characterization, another proof of the inequality by Audenaert [1] appeared. In this chapter, based on Audenaert's proof, we give a simple proof of the characterization given in [4].

We denote by $M_n(\mathbb{C})$ the vector space of all $n \times n$ complex matrices with the inner product $\langle X, Y \rangle = \mathrm{tr}(Y^*X)$, where $\mathrm{tr}(X)$ denotes the trace of X and Y^* is the conjugate transpose of Y. The Frobenius norm of X is defined by $\|X\|_F \equiv \sqrt{\langle X, X \rangle}$. We write $X \perp Y$ when X and Y are orthogonal. The identity matrix of order n is denoted by I_n, and we use O to denote zero matrices of appropriate order. We also use the direct sum notation $A \oplus B = \begin{bmatrix} A & O \\ O & B \end{bmatrix}$.

2.2 Main results

Theorem 2.1 **([4])** *Let $n > 1$ and $X, Y \in M_n(\mathbb{C})\backslash\{O\}$. Then X and Y satisfy (2.1) with equality if and only if $X \perp Y$ and there exists a unitary matrix U such that*

$$U^*XU,\ U^*YU \in \left\{ \begin{bmatrix} x & y \\ z & -x \end{bmatrix} \oplus O \ :\ x, y, z \in \mathbb{C} \right\}.$$

In order to give a simple proof of this theorem, we need the following lemmas.

Lemma 2.1 *Let $n \geq 2$, $x_1 \leq \cdots \leq x_n$, $p_j \geq 0$ for $j = 1, \ldots, n$, and $\sum_{j=1}^{n} p_j = 1$. Then*

$$\sum_{j=1}^{n} p_j x_j^2 - \Big(\sum_{j=1}^{n} p_j x_j \Big)^2 \leq \frac{1}{2} \sum_{j=1}^{n} x_j^2.$$

Moreover, assuming $x_1 < x_n$, the equality holds if and only if $p_1 = p_n = \frac{1}{2}$, $x_1 = -x_n$, and $x_2 = x_3 = \cdots = x_{n-1} = 0$.

Proof: Let $\mu = \sum_{j=1}^{n} p_j x_j$. Then

$$\begin{aligned}
\sum_{j=1}^{n} p_j x_j^2 - \Big(\sum_{j=1}^{n} p_j x_j \Big)^2 &= (\mu - x_1)(x_n - \mu) + \sum_{j=1}^{n} p_j (x_j - x_1)(x_j - x_n) \\
&\leq (\mu - x_1)(x_n - \mu) && (2.2) \\
&\leq \left[\frac{1}{2}(\mu - x_1 + x_n - \mu) \right]^2 && (2.3) \\
&= \frac{1}{2}(x_1^2 + x_n^2) - \frac{1}{4}(x_1 + x_n)^2 \\
&\leq \frac{1}{2}(x_1^2 + x_n^2) && (2.4) \\
&\leq \frac{1}{2} \sum_{j=1}^{n} x_j^2. && (2.5)
\end{aligned}$$

We now consider the equality. The sufficiency can be checked directly. Now assume that the equality holds. Then (2.2)–(2.5) hold in equality. From the equality in (2.5), we have

$$x_2 = x_3 = \cdots = x_{n-1} = 0.$$

From the equality in (2.4), we obtain $x_1 = -x_n$. Then, the equality in (2.2) implies $p_i = 0$ for $i \neq 1$, n, and consequently from the equality of (2.3), we get $p_1 = p_n = \frac{1}{2}$. □

Lemma 2.2 *Let $A \in M_n(\mathbb{C})\backslash\{O\}$ be a Hermitian matrix and $J \in M_n(\mathbb{C})$ be a density matrix, i.e., J is positive semi-definite and* $\operatorname{tr}(J) = 1$, *then*

$$\operatorname{tr}(JA^2) - [\operatorname{tr}(JA)]^2 \leq \frac{\|A\|_F^2}{2}.$$

Moreover, the equality holds if and only if there exists a unitary matrix V such that

$$V^*AV = \begin{bmatrix} a & 0 \\ 0 & -a \end{bmatrix} \oplus O \quad \textit{and} \quad V^*JV = \begin{bmatrix} \frac{1}{2} & p \\ \bar{p} & \frac{1}{2} \end{bmatrix} \oplus O,$$

where a is a real number, $p \in \mathbb{C}$, and $\bar{p}$ is the complex conjugate of p.

Proof: Since A is Hermitian, there is a unitary matrix U such that $U^*AU = D$, where $D = \operatorname{diag}(a_1, a_2, \ldots, a_n)$ with all a_i $(i = 1, \ldots, n)$ being real numbers. It follows that

$$\|A\|_F^2 = \|D\|_F^2 = \sum_{i=1}^{n} a_i^2,$$

and

$$\begin{aligned}\operatorname{tr}(JA^2) - [\operatorname{tr}(JA)]^2 &= \operatorname{tr}(JUD^2U^*) - [\operatorname{tr}(JUDU^*)]^2 \\ &= \operatorname{tr}(U^*JUD^2) - [\operatorname{tr}(U^*JUD)]^2 \\ &= \sum_{j=1}^{n} p_j a_j^2 - \Big(\sum_{j=1}^{n} p_j a_j\Big)^2,\end{aligned}$$

where $p_1, \ldots, p_n$ are the diagonal entries of U^*JU. Since J is a density matrix, the diagonal entries $p_1, \ldots, p_n$ are nonnegative and their sum is 1. We have by Lemma 2.1,

$$\operatorname{tr}(JA^2) - [\operatorname{tr}(JA)]^2 \leq \sum_{j=1}^{n} \frac{a_j^2}{2} = \frac{\|A\|_F^2}{2}.$$

Also, from Lemma 2.1, we know that the equality holds if and only if there exists a permutation matrix P such that $P^T(U^*AU)P = \operatorname{diag}(a, -a, 0, \ldots, 0)$ and

$$P^T(U^*JU)P = \begin{bmatrix} H & \times \\ \times & K \end{bmatrix},$$

where $H = \begin{bmatrix} \frac{1}{2} & \times \\ \times & \frac{1}{2} \end{bmatrix}_{2\times 2}$ and K has zero diagonal entries. Since $P^T(U^*JU)P$ is positive semi-definite, a zero diagonal entry implies that all entries in the same column and row are also zero. Hence

$$P^TU^*JUP = \begin{bmatrix} \frac{1}{2} & p \\ \bar{p} & \frac{1}{2} \end{bmatrix} \oplus O.$$

The result follows by letting $V = UP$. □

Lemma 2.3 *If $X \in M_n(\mathbb{C})\backslash\{O\}$ and $\operatorname{tr}(X) = 0$, then the density matrix $J = \frac{XX^*+X^*X}{2\|X\|_F^2}$ is at least of rank 2.*

Proof: Since X^*X is positive semi-definite, we have

$$\operatorname{rank}(J) \geq \operatorname{rank}(XX^*) = \operatorname{rank}(X).$$

It suffices to prove the case when $\operatorname{rank}(X) = 1$. Suppose that $\operatorname{rank}(X) = 1$. As $\operatorname{tr}(X) = 0$, there is a unitary matrix U such that

$$U^*XU = \|X\|_F E_{21},$$

where E_{jk} is the matrix with all entries being zero except the (j,k)th entry which equals 1. Consequently,

$$\begin{aligned} XX^* + X^*X &= \|X\|_F^2 U E_{21} E_{12} U^* + \|X\|_F^2 U E_{12} E_{21} U^* \\ &= \|X\|_F^2 U(E_{22} + E_{11})U^*. \end{aligned}$$

Thus, $\operatorname{rank}(J) = \operatorname{rank}(E_{22} + E_{11}) = 2$ which is the desired result. □

Proof of Theorem 2.1: The sufficiency can be checked directly. We just need to show the necessity. The following proof of inequality (2.1) is due to Audenaert [1]. Let $J = \frac{XX^*+X^*X}{2\|X\|_F^2}$, we have

$$\begin{aligned} &\|XY - YX\|_F^2 \\ &= \operatorname{tr}((X^*X + XX^*)(Y^*Y + YY^*)) - \|X^*Y + YX^*\|_F^2 \\ &\leq \operatorname{tr}((X^*X + XX^*)(Y^*Y + YY^*)) - \frac{|\operatorname{tr}((X^*Y + YX^*)X)|^2}{\|X\|_F^2} \qquad (2.6) \\ &= 4\|X\|_F^2 \left[\operatorname{tr}\left(\frac{J(Y^*Y + YY^*)}{2}\right) - |\operatorname{tr}(JY)|^2\right]. \end{aligned}$$

Note that inequality (2.6) follows from the Cauchy-Schwarz inequality. Consider the Cartesian decomposition of $Y = A + \mathbf{i}B$, where A and B are Hermitian matrices, and $\mathbf{i} \equiv \sqrt{-1}$. Then by Lemma 2.2,

$$\begin{aligned} \|XY - YX\|_F^2 &\leq 4\|X\|_F^2 \left[\operatorname{tr}(J(A^2 + B^2)) - [\operatorname{tr}(JA)]^2 - [\operatorname{tr}(JB)]^2\right] \\ &= 4\|X\|_F^2 \left[\left(\operatorname{tr}(JA^2) - [\operatorname{tr}(JA)]^2\right) + \left(\operatorname{tr}(JB^2) - [\operatorname{tr}(JB)]^2\right)\right] \\ &\leq 4\|X\|_F^2 \left(\frac{\|A\|_F^2}{2} + \frac{\|B\|_F^2}{2}\right) \qquad (2.7) \\ &= 2\|X\|_F^2 \|Y\|_F^2. \end{aligned}$$

Note that (2.7) follows from Lemma 2.2.

Now assume that (2.1) holds with equality. Then (2.7) holds with equality. We assume that both A and B are nonzero. Otherwise, the proof can be modified easily. By Lemma 2.2, there exist unitary matrices U and V such that

$$U^*AU = \begin{bmatrix} a & 0 \\ 0 & -a \end{bmatrix} \oplus O \quad \text{and} \quad U^*JU = J_A \oplus O, \qquad (2.8)$$

$$V^*BV = \begin{bmatrix} b & 0 \\ 0 & -b \end{bmatrix} \oplus O \quad \text{and} \quad V^*JV = J_B \oplus O, \qquad (2.9)$$

where both J_A and J_B are in the form of $\begin{bmatrix} \frac{1}{2} & p \\ \bar{p} & \frac{1}{2} \end{bmatrix}$. Let

$$U^*V = W = \begin{bmatrix} W_{11} & W_{12} \\ W_{21} & W_{22} \end{bmatrix},$$

where $W_{11} \in M_2(\mathbb{C})$. Then from the two different expressions of J, we have

$$U(J_A \oplus O)U^* = V(J_B \oplus O)V^*$$

or equivalently $(J_A \oplus O)U^*V = U^*V(J_B \oplus O)$, which gives

$$\begin{bmatrix} J_A W_{11} & J_A W_{12} \\ O & O \end{bmatrix} = \begin{bmatrix} W_{11}J_B & O \\ W_{21}J_B & O \end{bmatrix}. \tag{2.10}$$

From (2.8) and (2.9), we see that $\mathrm{tr}(Y) = \mathrm{tr}(A) + \mathbf{i} \cdot \mathrm{tr}(B) = 0$. Because of the symmetry of X and Y in (2.1) and its corresponding equality, we have $\mathrm{tr}(X) = 0$. Then by Lemma 2.3, $\mathrm{rank}(J) = 2$ and hence J_A and J_B are invertible. As a consequence, from (2.10), we can deduce that W_{12} and W_{21} are zero matrices and so $W = W_{11} \oplus W_{22}$. Then

$$\begin{aligned} U^*YU &= U^*(A + \mathbf{i}B)U \\ &= \left(\begin{bmatrix} a & 0 \\ 0 & -a \end{bmatrix} \oplus O\right) + \mathbf{i}W\left(\begin{bmatrix} b & 0 \\ 0 & -b \end{bmatrix} \oplus O\right)W^* \\ &= \left(\begin{bmatrix} a & 0 \\ 0 & -a \end{bmatrix} \oplus O\right) + \mathbf{i}\left(W_{11}\begin{bmatrix} b & 0 \\ 0 & -b \end{bmatrix} W_{11}^*\right) \oplus O \\ &= \begin{bmatrix} y_{11} & y_{12} \\ y_{21} & -y_{11} \end{bmatrix} \oplus O. \end{aligned}$$

It remains to characterize X. Since

$$J = \frac{XX^* + X^*X}{2\|X\|_F^2} \quad \text{and} \quad U^*JU = J_A \oplus O,$$

it follows that

$$[U^*(XX^* + X^*X)U]_{j,j} = 0,$$

for $j = 3, \ldots, n$. Hence,

$$[(U^*XU)(U^*XU)^*]_{j,j} = 0, \qquad [(U^*XU)^*(U^*XU)]_{j,j} = 0,$$

for $j = 3, \ldots, n$. This implies that all entries in the jth row and the jth column of U^*XU are equal to zero for $j = 3, \ldots, n$, i.e.,

$$U^*XU = \begin{bmatrix} x_{11} & x_{12} \\ x_{21} & -x_{11} \end{bmatrix} \oplus O.$$

Finally, for (2.6) to hold in equality, we have

$$XY^* + Y^*X = \alpha X$$

for some $\alpha \in \mathbb{C}$. As $\mathrm{tr}(X) = 0$, we obtain $\mathrm{tr}(Y^*X) = 0$, i.e., $X \perp Y$. □

References for Chapter 2

[1] K. Audenaert, *Variance bounds, with an application to norm bounds for commutators*, Linear Algebra Appl., 432 (2010), pp. 1126–1143.

[2] A. Böttcher and D. Wenzel, *How big can the commutator of two matrices be and how big is it typically?*, Linear Algebra Appl., 403 (2005), pp. 216–288.

[3] A. Böttcher and D. Wenzel, *The Frobenius norm and the commutator*, Linear Algebra Appl., 429 (2008), pp. 1864–1885.

[4] C.-M. Cheng, S.-W. Vong, and D. Wenzel, *Commutators with maximal Frobenius norm*, Linear Algebra Appl., 432 (2010), pp. 292–306.

[5] L. László, *Proof of Böttcher and Wenzel's conjecture on commutator norms for 3-by-3 matrices*, Linear Algebra Appl., 422 (2007), pp. 659–663.

[6] Z.-Q. Lu, *Proof of the normal scalar curvature conjecture*, 22 November 2007. Available from: <arXiv:0711.3510v1[math.DG]>.

[7] S.-W. Vong and X.-Q. Jin, *Proof of Böttcher and Wenzel's conjecture*, Oper. Matrices, 2 (2008), pp. 435–442.

Chapter 3

On High-dimensional Markov Chain Models for Categorical Data Sequences with Applications

Wai-ki Ching [1] Dong-mei Zhu [2]

Abstract

In this chapter, we give a brief review on some recent models and results in high-dimensional Markov chain models for categorical data sequences. Categorical data sequences occur in many areas including bioinformatics, operations research, data mining, financial data modeling, and so on. The construction of mathematical models for categorical data sequences is important in the sense that it can help in the discovery of rules, testing hypothesis, and making effective predictions. Here we introduce a number of high-dimensional Markov chain models with applications. The high-dimensional models include high-order Markov chain models, multivariate Markov chain models, and high-order multivariate Markov chain models. Efficient estimation methods for the model parameters are also discussed.

3.1 Introduction

Categorical data sequences have found their applications in many real world problems. In this chapter, we give a brief survey on high-dimensional Markov chain models for modeling categorical data sequences with applications. The construction of mathematical models

[1]Advanced Modeling and Applied Computing Laboratory, Department of Mathematics, The University of Hong Kong, Pokfulam Road, Hong Kong (email: wching@hkusua.hku.hk).
Dedicated to the 20th anniversary of TOTS founded by Prof. Raymond H. Chan. Research supported in part by HKRGC Grant No. 7017/07P, HKUCRGC Grants, HKU Strategy Research Theme fund on Computational Sciences, Hung Hing Ying Physical Research Sciences Research Grant, National Natural Science Foundation of China Grant No. 10971075, and Guangdong Provincial Natural Science Grant No. 9151063101000021.
[2]Advanced Modeling and Applied Computing Laboratory, Department of Mathematics, The University of Hong Kong, Pokfulam Road, Hong Kong (email: dongmeizhu86@gmail.com).

for categorical data sequences is important in the sense that it can help in the discovery of rules, testing hypothesis, and making effective predictions. Here a data sequence X is represented by using a time series notation

$$\{X_1, X_2, \ldots, X_{T-1}, X_T\}, \tag{3.1}$$

where T is the length of the data sequence and each X_i is in certain domain $\mathcal{M}$ or state-space. For example, we may have

$$\mathcal{M} = \{1, 2, \ldots, m\} \tag{3.2}$$

with an integer m. The domain is associated with a defined semantic and a data type. A numeric domain consists of real numbers. A domain $\mathcal{M}$ is defined as categorical if it is finite and can be unordered, which means that for any $x, y \in M$, we have either $x = y$ or $x \neq y$. Using the state-space $\mathcal{M}$, the categorical data sequence in (3.1) can be represented as a sequence of unit vectors $\{\mathbf{x}_1, \mathbf{x}_2, \ldots, \mathbf{x}_{T-1}, \mathbf{x}_T\}$, where $\mathbf{x}_i = \mathbf{e}_k$ if $X_i = k$, and $\mathbf{e}_k$ is the unit vector with the kth entry being equal to one. This representation is useful when we use transition probability matrix to describe the dynamics and relations among the sequences.

For categorical data sequences, Markov chain models seem to be an effective tool [29]. Markov chain models occur frequently in many real world applications, such as queueing networks [3, 5], tele-communication networks [5, 6], inventory control in a production system [7, 8], re-manufacturing systems [36], bioinformatics [11, 12, 34, 37], financial modeling [17, 18, 19], and even in modeling categorical data sequences [14, 26]. We consider a simple Markov process X_t taking values in a finite state-space $\mathcal{M}$ over discrete time intervals $t = 1, 2, \ldots$. The probability distribution of X_{t+1} (at time $t+1$) only depends on the state X_t (at time t). We note that a Markov chain process is characterized by its transition probability matrix. Suppose that an observed sequence of states $\{X_t\}$ is given, then the transition probabilities among the states can be estimated by just counting the transition frequencies among the states and followed by a normalization [11]. Furthermore, it is well known that, for an irreducible Markov chain, the existence of the system steady-state distribution is guaranteed by the famous Perron-Frobenius theorem [2]. The steady-state distribution is important in the sense that it gives the long-run behavior (probability distribution) of the system. Therefore, by making use of the steady-state distribution, many useful and important system performance indicators can be obtained. Because of these important and nice properties, Markov chain models have been widely used in many different disciplines. However, in some situations, further improvement can be made by extending a simple Markov chain model to a high-order Markov chain model or a multivariate Markov chain model.

In a simple Markov process X_t taking values in the state-space $\mathcal{M}$ over the discrete time intervals $t = 1, 2, \ldots$, the probability distribution of X_{t+1} only depends on X_t. In some circumstances, X_{t+1} may also depend on the previous observations $\{X_{t-1}, X_{t-2}, \ldots, X_{t-n+1}\}$ as well. In this situation, it is natural for one to consider a high-order Markov chain model. In fact, high-order Markov chain models have been shown to be effective for modeling categorical data sequences [30, 31]. In a conventional model of an nth-order Markov chain model with m categories (states), there are $O(m^{n+1})$ parameters to be estimated. Thus one can see that the number of parameters increases exponentially with respect to the order of the model. In this chapter, we give a brief survey on some high-order Markov chain models. The models only involve reasonably many parameters and are able to capture the high-order correlation in categorical data sequences.

Apart from high-order Markov chain models, there are also situations that one would like to consider a number of Markov chains $\{X_{t,i}\}_{i=1}^{s}$ at the same time. The advantage of this approach, of course, is to make use of the extra information about the states of other chains at time t. It is usual that $X_{t+1,i}$, the state of the ith chain at time $t+1$, depends not only on $X_{t,i}$, but also on $\{X_{t,1}, \ldots, X_{t,i-1}, X_{t,i+1}, \ldots, X_{t,s}\}$, i.e., the states of other $(s-1)$ chains at time t. This will then result in a multivariate Markov chain model. We note that in a conventional model of multivariate Markov chain (a high-dimensional Markov chain) of s chains, each chain has the same set of m states, and the total number of parameters is $O(m^s)$. Thus the number of parameters increases exponentially with respect to the number of chains s. Therefore, there is a need to develop new multivariate Markov chain models which can capture both the intra- and inter-transition among the chains with reasonable number of model parameters.

The rest of the chapter is structured as follows. In section 3.2, we give a survey on the high-order Markov chain models. Both numerical examples and practical problems are discussed. In section 3.3, we give a survey on multivariate Markov chain models. We present the estimation methods with both numerical examples and practical problems to demonstrate the effectiveness of the models. In section 3.4, we present some variation of the models including the high-order multivariate Markov chain model. Finally concluding remarks are given in section 3.5 to address further research issues.

3.2 High-order Markov chain models

We give a brief survey on some high-order Markov chain models for categorical data sequences. To facilitate our discussion, we assume that each data X_t in a categorical data sequence takes values in $\mathcal{M}$ (c.f. (3.2)), i.e., it has m possible categories (states). We remark that there are $O(m^{n+1})$ model parameters in a conventional nth-order Markov chain model. Raftery [30] proposed a high-order Markov chain model which only involves one additional parameter for each extra lag. The model can be described as follows:

$$P(X_t = k_0 \mid X_{t-1} = k_1, \ldots, X_{t-n} = k_n) = \sum_{i=1}^{n} \lambda_i q_{k_0 k_i}, \tag{3.3}$$

where

$$\sum_{i=1}^{n} \lambda_i = 1$$

and $Q = [q_{ij}]$ is a transition matrix with column sums equal to one, such that

$$0 \leq \sum_{i=1}^{n} \lambda_i q_{k_0 k_i} \leq 1, \quad k_0, k_i \in \mathcal{M}. \tag{3.4}$$

The constraints in (3.4) are used to guarantee that the right-hand-side of (3.3) is a probability distribution. The total number of independent parameters in his model is $O(n+m^2)$. Raftery showed that (3.3) is analogous to the standard AR(n) model in the sense that each additional lag, after the first, is specified by a single parameter and the auto-correlations satisfy a system of linear equations similar to the Yule-Walker equations. Regarding the model parameters $q_{k_0 k_i}, \lambda_i$, they can be estimated numerically by maximizing the log-likelihood of (3.3) subject to the constraints in (3.4). The estimation method involves solving a highly nonlinear optimization problem and the proposed numerical method neither guarantees convergence nor a global maximum.

Later Ching et al. [8, 9] proposed a new model based on the Raftery's model [30] by allowing Q to be varied with different lags. An efficient method is developed based on linear programming to estimate the model parameters [10]. We note that (3.3) can be re-written as

$$\mathbf{x}_{t+n+1} = \sum_{i=1}^{n} \lambda_i Q \mathbf{x}_{t+n+1-i}, \tag{3.5}$$

where $\mathbf{x}_{t+n+1-i}$ is the probability distribution of the states at time $t+n+1-i$. Using (3.5) and the fact that Q is a transition probability matrix, each entry of $\mathbf{x}_{t+n+1}$ is between 0 and 1, and the sum of all entries is also equal to one. We note that in the Raftery's model the weightings λ_i are not necessarily nonnegative. Therefore the additional constraints (3.4) have to be added to guarantee that $\mathbf{x}_{t+n+1}$ is the probability distribution of the states. Ching et al. [7] generalize the Raftery's model in (3.5) as follows:

$$\mathbf{x}_{t+n+1} = \sum_{i=1}^{n} \lambda_i Q_i \mathbf{x}_{t+n+1-i}. \tag{3.6}$$

The total number of independent parameters in the new model is $O(nm^2)$. The weightings λ_i are nonnegative such that

$$\sum_{i=1}^{n} \lambda_i = 1.$$

We remark that if

$$Q_1 = Q_2 = \cdots = Q_n,$$

then (3.6) is reduced to the Raftery model in (3.5). In fact, in the model (3.6), it is assumed that $\mathbf{x}_{t+n+1}$ depends on $\mathbf{x}_{t+i}$ ($i = 1, 2, \ldots, n$) via the matrices Q_i and the weightings λ_i. One may relate Q_i to the ith step transition matrix of the process and this idea is used to estimate Q_i. Before introducing their estimation method for the model parameters, we first present an important property of the new proposed model in the following proposition. The result indicates that the proposed model is a stationary model under some reasonable conditions.

Proposition 3.1 [9] *If Q_n is irreducible and $\lambda_n > 0$ such that*

$$0 \le \lambda_i \le 1 \quad \textit{and} \quad \sum_{i=1}^{n} \lambda_i = 1,$$

then the model in (3.6) has a stationary distribution $\bar{\mathbf{x}}$ when $t \to \infty$ independent of the initial state vectors $\mathbf{x}_0, \mathbf{x}_1, \ldots, \mathbf{x}_{n-1}$. The steady-state distribution $\bar{\mathbf{x}}$ is also the unique solution of the following linear system of equations:

$$\Big(I - \sum_{i=1}^{n} \lambda_i Q_i\Big)\bar{\mathbf{x}} = \mathbf{0} \quad \text{and} \quad \mathbf{1}^T \bar{\mathbf{x}} = 1,$$

where I is the m-by-m identity matrix and $\mathbf{1}$ is an m-vector of ones.

3.2.1 Parameters estimations for high-order Markov chain models

We introduce some efficient estimation methods for Q_i and λ_i for $i = 1, 2, \ldots, n$. To estimate Q_i, we regard Q_i as the ith step transition matrix of the categorical data sequence $\{X_t\}$. Suppose that an observed categorical data sequence $\{X_t\}_{t=1}^{T}$ is given, we can count the transition frequency $f_{jk}^{(i)}$ in the sequence from state k to state j in the ith step. Hence one can construct the ith step transition matrix for the observed sequence $\{X_t\}_{t=1}^{T}$ as follows:

$$F^{(i)} = \begin{bmatrix} f_{11}^{(i)} & \cdots & \cdots & f_{m1}^{(i)} \\ f_{12}^{(i)} & \cdots & \cdots & f_{m2}^{(i)} \\ \vdots & \vdots & \vdots & \vdots \\ f_{1m}^{(i)} & \cdots & \cdots & f_{mm}^{(i)} \end{bmatrix}.$$

From $F^{(i)}$, we obtain the estimates for $Q_i = [q_{kj}^{(i)}]$ as follows:

$$\hat{Q}_i = \begin{bmatrix} \hat{q}_{11}^{(i)} & \cdots & \cdots & \hat{q}_{m1}^{(i)} \\ \hat{q}_{12}^{(i)} & \cdots & \cdots & \hat{q}_{m2}^{(i)} \\ \vdots & \vdots & \vdots & \vdots \\ \hat{q}_{1m}^{(i)} & \cdots & \cdots & \hat{q}_{mm}^{(i)} \end{bmatrix},$$

where

$$\hat{q}_{kj}^{(i)} = \begin{cases} \dfrac{f_{kj}^{(i)}}{\sum_{k=1}^{m} f_{kj}^{(i)}}, & \text{if } \sum_{k=1}^{m} f_{kj}^{(i)} \neq 0, \\ 0, & \text{otherwise.} \end{cases}$$

The computational cost for the construction of $F^{(i)}$ is of $O(T^2)$. Hence the total computational cost for the construction of $\{F^{(i)}\}_{i=1}^{n}$ is of $O(nT^2)$.

Linear programming formulation for estimation of λ_i

Proposition 3.1 gives a sufficient condition for the model to be stationary, i.e., $\mathbf{x}_t$ converges to a stationary distribution $\bar{\mathbf{x}}$ ($\mathbf{x}_t \to \bar{\mathbf{x}}$ as $t \to \infty$). Then $\bar{\mathbf{x}}$ can be estimated from the sequence $\{X_t\}$ by calculating the proportion of the occurrence of each state in the observed data sequence and we denote it by $\hat{\mathbf{x}}$. Hence one would expect the observed distribution to be close to the theoretical one, i.e.,

$$\sum_{i=1}^{n} \lambda_i \hat{Q}_i \hat{\mathbf{x}} \approx \hat{\mathbf{x}}.$$

This suggests that the model parameters $\lambda = \{\lambda_1, \ldots, \lambda_n\}$ can be obtained by minimizing the gap. We remark that many norms, such as $\|\cdot\|_2$ and $\|\cdot\|_1$, can also be adopted for the captured problem. The former will result in a quadratic programming problem, while the latter will result in a linear programming problem (see, for instance, [22, pp.221-226]). Moreover, it is also known that, in approximating data by a linear function [22, pp.220], $\|\cdot\|_1$ always gives the most robust answer and $\|\cdot\|_\infty$ avoids gross discrepancies with the data. If the errors are known to be normally distributed, then $\|\cdot\|_2$ is the best.

In the following, as a demonstration, we adopt $\|\cdot\|_\infty$ as the norm and we consider the following optimization problem (a min-max problem):

$$\min_{\lambda} \max_{k} \left| \left[\sum_{i=1}^{n} \lambda_i \hat{Q}_i \hat{\mathbf{x}} - \hat{\mathbf{x}} \right]_k \right|$$

subject to

$$\sum_{i=1}^{n} \lambda_i = 1, \quad \text{and} \quad \lambda_i \geq 0, \quad \forall i,$$

where $[\cdot]_k$ denotes the kth entry of the vector. The constraints in the optimization problem guarantee the existence of the stationary distribution $\bar{\mathbf{x}}$ which is estimated by $\hat{\mathbf{x}}$. We further observe that the above optimization problem can be re-formulated as a linear programming problem [22]:

$$\min_{\lambda} w$$

subject to

$$\begin{bmatrix} w \\ w \\ \vdots \\ w \end{bmatrix} \geq \hat{\mathbf{x}} - \left[\hat{Q}_1 \hat{\mathbf{x}} \mid \hat{Q}_2 \hat{\mathbf{x}} \mid \ldots \mid \hat{Q}_n \hat{\mathbf{x}} \right] \begin{bmatrix} \lambda_1 \\ \lambda_2 \\ \vdots \\ \lambda_n \end{bmatrix},$$

$$\begin{bmatrix} w \\ w \\ \vdots \\ w \end{bmatrix} \geq -\hat{\mathbf{x}} + \left[\hat{Q}_1 \hat{\mathbf{x}} \mid \hat{Q}_2 \hat{\mathbf{x}} \mid \ldots \mid \hat{Q}_n \hat{\mathbf{x}} \right] \begin{bmatrix} \lambda_1 \\ \lambda_2 \\ \vdots \\ \lambda_n \end{bmatrix},$$

$$w \geq 0, \quad \sum_{i=1}^{n} \lambda_i = 1, \quad \text{and} \quad \lambda_i \geq 0, \quad \forall i.$$

One can solve the above linear programming problem efficiently by using EXCEL [10] and obtain the model parameters λ.

A numerical demonstration

We demonstrate the estimation method by a simple example [9]. We consider a sequence $\{X_t\}$ of three states ($m = 3$) given by

$$\{1, 1, 2, 2, 1, 3, 2, 1, 2, 3, 1, 2, 3, 1, 2, 3, 1, 2, 1, 2\}. \tag{3.7}$$

The sequence $\{X_t\}$ can be written in vector form

$$\mathbf{x}_1 = (1, 0, 0)^T, \quad \mathbf{x}_2 = (1, 0, 0)^T, \quad \ldots, \quad \mathbf{x}_{19} = (1, 0, 0)^T, \quad \mathbf{x}_{20} = (0, 1, 0)^T.$$

We consider a second-order ($n = 2$) model. Then from (3.7), we have the transition frequency matrices

$$F^{(1)} = \begin{bmatrix} 1 & 3 & 3 \\ 6 & 1 & 1 \\ 1 & 3 & 0 \end{bmatrix} \quad \text{and} \quad F^{(2)} = \begin{bmatrix} 1 & 4 & 1 \\ 3 & 2 & 3 \\ 3 & 1 & 0 \end{bmatrix}. \tag{3.8}$$

From (3.8), we have the ith step transition matrices ($i = 1, 2$) as follows:

$$\hat{Q}_1 = \begin{bmatrix} 1/8 & 3/7 & 3/4 \\ 3/4 & 1/7 & 1/4 \\ 1/8 & 3/7 & 0 \end{bmatrix}, \qquad \hat{Q}_2 = \begin{bmatrix} 1/7 & 4/7 & 1/4 \\ 3/7 & 2/7 & 3/4 \\ 3/7 & 1/7 & 0 \end{bmatrix},$$

and

$$\hat{\mathbf{x}} = \left(\frac{2}{5}, \frac{2}{5}, \frac{1}{5}\right)^T.$$

Hence we have

$$\hat{Q}_1\hat{\mathbf{x}} = \left(\frac{13}{35}, \frac{57}{140}, \frac{31}{140}\right)^T \quad \text{and} \quad \hat{Q}_2\hat{\mathbf{x}} = \left(\frac{47}{140}, \frac{61}{140}, \frac{8}{35}\right)^T.$$

To estimate λ_i, we consider the optimization problem:

$$\min_{\lambda_1, \lambda_2} w$$

subject to

$$\begin{cases} w \geq \frac{2}{5} - \frac{13}{35}\lambda_1 - \frac{47}{140}\lambda_2, \\ w \geq -\frac{2}{5} + \frac{13}{35}\lambda_1 + \frac{47}{140}\lambda_2, \\ w \geq \frac{2}{5} - \frac{57}{140}\lambda_1 - \frac{61}{140}\lambda_2, \\ w \geq -\frac{2}{5} + \frac{57}{140}\lambda_1 + \frac{61}{140}\lambda_2, \\ w \geq \frac{1}{5} - \frac{31}{140}\lambda_1 - \frac{8}{35}\lambda_2, \\ w \geq -\frac{1}{5} + \frac{31}{140}\lambda_1 + \frac{8}{35}\lambda_2, \\ w \geq 0, \quad \lambda_1 + \lambda_2 = 1, \quad \lambda_1, \lambda_2 \geq 0. \end{cases}$$

The optimal solution is $(\lambda_1^*, \lambda_2^*, w^*) = (1, 0, 0.0286)$, and we have the following high-order Markov model

$$\mathbf{x}_{t+1} = \hat{Q}_1\mathbf{x}_t.$$

This means the optimal model is of first-order, even if we assume a second-order setting.

3.2.2 Some practical applications

We present some applications of the high-order Markov chain model. The applications include: (i) DNA sequences, (ii) Web prediction, and (iii) inventory control.

(i) DNA sequences

A DNA sequence can be regarded as a categorical data sequence of four possible categories: A, C, T, and G. To determine whether certain short DNA sequence occurred more often than would be expected by chance, Avery [1] examined the Markovian structure of introns from several other genes in mice. Here the model is applied to the introns from the mouse

αA-crystallin gene (see, for instance, [31]). In the following, a comparison between Ching's second-order model and Raftery's second-order model is given. The results are reported in Table 3.1 below (taken from [9]). The comparison is made with different grouping of states as suggested in [31]. We observe that the Ching's model gives better predictions when compared with the Raftery's model.

Table 3.1: Prediction Accuracy in the DNA Sequence.

	2-state model	3-state model	4-state model
Ching's High-order Chain Model	57%	49%	33%
Raftery's High-order Chain Model	57%	47%	31%
Random Chosen	50%	33%	25%

(ii) Web prediction

The model can be applied to Web prediction [35]. As we know that the Internet provides a rich platform for users to surf for information. One way to assist the users with their needs for useful information is to predict a user's next target and use the prediction as a recommendation. Such a recommendation system relies on a reliable prediction model to make inferences on users' interests. Examples can be found in WebWatcher system [24] and Letzia system [28]. In [9], a high-order Markov model was built on a Web server log file in the following manner. A Webpage visited by a user is regarded as a state and each session is then a categorical data sequence. The high-order Markov chain model can capture the useful information about the behavior of a user and hence gives a reliable recommendation system.

(iii) Inventory control

The high-order Markov chain model has been applied to a generalized Newsboy's problem [8]. It is about a production problem faced by a large soft-drink company in Hong Kong. There are product categories due to different sales volume. All products can be labeled as either very high sales volume (state 1), higher sales volume (state 2), standard sales volume (state 3), lower sales volume (state 4), or very low sales volume (state 5). The labels are useful from both marketing and production planning points of view. They assume the following penalty cost matrix for three products (A, B, and C):

$$C = \begin{bmatrix} 0 & 100 & 300 & 700 & 1500 \\ 100 & 0 & 100 & 300 & 700 \\ 300 & 100 & 0 & 100 & 300 \\ 700 & 300 & 100 & 0 & 100 \\ 1500 & 700 & 300 & 100 & 0 \end{bmatrix},$$

where $[C]_{ij}$ is the cost when the production plan is for sales volume of state i, but the actual sales volume is state j. For the high-order Markov chain model, again they consider the third-order ($n = 3$) model. The model is then compared with the first order Markov model and the stationary model for the three product demand sequences. The results are reported in Table 3.2 (taken from [8]). Clearly the high-order Markov chain model performs the best.

Table 3.2: The Optimal Costs of the Three Different Models.

	Product A	Product B	Product C
Ching's High-order Markov Chain Model ($n = 3$)	11200	9300	10800
First-order Markov Chain Model	27600	18900	11100
Stationary Model	31900	18900	16300

3.3 Multivariate Markov chain models

We now consider modeling multiple categorical data sequences using Markov chain model. We assume that there are s sequences and each sequence has m possible common states. If we model the joint sequences by a conventional Markov chain model, then it has $O(m^s)$ states. The number of model parameters increases exponentially with respect to the number of the joint sequences. Ching et al. [7] present a multivariate Markov chain model which captures both the intra- and inter-transition probabilities among the sequences. The total number of parameters in the model is only $O(s^2m^2)$. They also present efficient methods to estimate the model parameters. Under some reasonable condition, the model can be shown to be a stationary one and the result is an analogy to the Perron-Frobenious Theorem for a nonnegative matrix (see, for instance, [2]).

Proposition 3.2 (Perron-Frobenius Theorem) *Let A be a nonnegative and irreducible square matrix of order m. Then*

(i) *A has a positive real eigenvalue γ equal to its spectral radius, i.e., $\gamma = \max_k |\rho_k(A)|$, where $\rho_k(A)$ denotes the kth eigenvalue of A.*

(ii) *There corresponds an eigenvector $\mathbf{z}$, its entries being real and positive, such that*

$$A\mathbf{z} = \gamma\mathbf{z}.$$

(iii) *γ is a simple eigenvalue of A.*

In the following, we introduce the multivariate Markov chain model discussed in [7] for multiple categorical sequences generated by similar sources or the same source. We assume that there are s categorical sequences and each has m possible states in the same state-space $\mathcal{M}$. Let $\mathbf{x}_t^{(k)}$ be the state vector of the kth sequence at time t. If the kth sequence is in state j at time t, then $\mathbf{x}_t^{(k)} = \mathbf{e}_j$. In the multivariate Markov chain model, the following relationships are assumed:

$$\mathbf{x}_{t+1}^{(j)} = \sum_{k=1}^{s} \lambda_{jk} P^{(jk)} \mathbf{x}_t^{(k)}, \quad \text{for} \quad j = 1, 2, \ldots, s,$$

where

$$\lambda_{jk} \geq 0, \quad \text{for} \quad 1 \leq j, k \leq s; \quad \text{and} \quad \sum_{k=1}^{s} \lambda_{jk} = 1, \quad \text{for} \quad j = 1, 2, \ldots, s.$$

The state probability distribution of the kth sequence at the $(t+1)$th step depends on the weighted average of $P^{(jk)}\mathbf{x}_t^{(k)}$. Here $P^{(jk)}$ is a transition probability matrix from the states in the kth sequence to the states in the jth sequence, and $\mathbf{x}_t^{(k)}$ is the state probability distribution of the kth sequence at the tth step. In matrix form, one can write

$$\mathbf{x}_{t+1} \equiv \begin{bmatrix} \mathbf{x}_{t+1}^{(1)} \\ \mathbf{x}_{t+1}^{(2)} \\ \vdots \\ \mathbf{x}_{t+1}^{(s)} \end{bmatrix} = \begin{bmatrix} \lambda_{11}P^{(11)} & \lambda_{12}P^{(12)} & \cdots & \lambda_{1s}P^{(1s)} \\ \lambda_{21}P^{(21)} & \lambda_{22}P^{(22)} & \cdots & \lambda_{2s}P^{(2s)} \\ \vdots & \vdots & \vdots & \vdots \\ \lambda_{s1}P^{(s1)} & \lambda_{s2}P^{(s2)} & \cdots & \lambda_{ss}P^{(ss)} \end{bmatrix} \begin{bmatrix} \mathbf{x}_t^{(1)} \\ \mathbf{x}_t^{(2)} \\ \vdots \\ \mathbf{x}_t^{(s)} \end{bmatrix} \equiv Q\mathbf{x}_t,$$

or

$$\mathbf{x}_{t+1} = Q\mathbf{x}_t. \tag{3.9}$$

Although the column sum of Q is not equal to one (the column sum of $P^{(jk)}$ is equal to one), we still have the following propositions [7].

Proposition 3.3 *If $\lambda_{jk} > 0$ for $1 \le j, k \le s$, then the matrix Q has an eigenvalue equal to one and the eigenvalues of Q have modulus less than or equal to 1.*

Proposition 3.4 *Suppose that $P^{(jk)}$ are irreducible and $\lambda_{jk} > 0$ for $1 \le j, k \le s$. Then there is a vector*

$$\mathbf{x} = \begin{bmatrix} \mathbf{x}^{(1)} \\ \mathbf{x}^{(2)} \\ \vdots \\ \mathbf{x}^{(s)} \end{bmatrix}$$

such that $\mathbf{x} = Q\mathbf{x}$ and

$$\sum_{i=1}^{m} [\mathbf{x}^{(j)}]_i = 1, \quad \text{for } 1 \le j \le s.$$

We note that the condition on $\lambda_{jk} > 0$ can be relaxed to a mild condition that Λ is an irreducible matrix, where $[\Lambda]_{j,k} = \lambda_{jk}$ [7]. The vector $\mathbf{x}$ is not a probability distribution vector, but $\mathbf{x}^{(j)}$ is a probability distribution vector. The above proposition suggests one possible way to estimate the model parameters λ_{ij}. The idea here is similar to the one in section 3.2 that we can find λ_{ij} by minimizing the gap $\|Q\hat{\mathbf{x}} - \hat{\mathbf{x}}\|$ under certain vector norm $\|\cdot\|$.

3.3.1 Parameters estimations for multivariate Markov chain models

We discuss efficient methods for estimating the parameters $P^{(jk)}$ and λ_{jk} for $j, k = 1, \ldots, s$ in the multivariate Markov chain model. Given the data sequences

$$\{X^{(1)}\}, \{X^{(2)}\}, \ldots, \{X^{(s)}\},$$

one can count the transition frequency $f_{i_j i_k}^{(jk)}$ from the state i_j in the sequence $\{X^{(j)}\}$ to the state i_k in the sequence $\{X^{(k)}\}$. Thus one can construct the transition frequency matrix for the sequences as follows:

$$F^{(jk)} = \begin{bmatrix} f_{11}^{(jk)} & \cdots & \cdots & f_{m1}^{(jk)} \\ f_{12}^{(jk)} & \cdots & \cdots & f_{m2}^{(jk)} \\ \vdots & \vdots & \vdots & \vdots \\ f_{1m}^{(jk)} & \cdots & \cdots & f_{mm}^{(jk)} \end{bmatrix}.$$

From $F^{(jk)}$, we obtain the estimates for $P^{(jk)}$ as follows:

$$\hat{P}^{(jk)} = \begin{bmatrix} \hat{p}_{11}^{(jk)} & \cdots & \cdots & \hat{p}_{m1}^{(jk)} \\ \hat{p}_{12}^{(jk)} & \cdots & \cdots & \hat{p}_{m2}^{(jk)} \\ \vdots & \vdots & \vdots & \vdots \\ \hat{p}_{1m}^{(jk)} & \cdots & \cdots & \hat{p}_{mm}^{(jk)} \end{bmatrix},$$

where

$$\hat{p}_{i_j i_k}^{(jk)} = \begin{cases} \dfrac{f_{i_j i_k}^{(jk)}}{\sum_{i_k=1}^{m} f_{i_j i_k}^{(jk)}}, & \text{if } \sum_{i_k=1}^{m} f_{i_j i_k}^{(jk)} \neq 0, \\ 0, & \text{otherwise.} \end{cases}$$

We know that under some conditions, the multivariate Markov chain has a steady-state distribution vector $\mathbf{x}$. The vector $\mathbf{x}$ can be estimated from the sequences by computing the proportion of the occurrence of each state in the sequence of each system and let us denote it by

$$\hat{\mathbf{x}} = \begin{bmatrix} \hat{\mathbf{x}}^{(1)} \\ \hat{\mathbf{x}}^{(2)} \\ \vdots \\ \hat{\mathbf{x}}^{(s)} \end{bmatrix}$$

Thus we expect $Q\hat{\mathbf{x}} \approx \hat{\mathbf{x}}$. This suggests one possible way to estimate the parameters $\lambda = \{\lambda_{jk}\}_{j,k=1}^{s}$ as follows. For demonstrative purpose, we choose $\|\cdot\|$ to be $\|\cdot\|_\infty$, then for each $j = 1, 2, \ldots, s$, we consider the following optimization problem:

$$\min_{\lambda} \max_{i} \left| \left[\sum_{k=1}^{s} \lambda_{jk} \hat{P}^{(jk)} \hat{\mathbf{x}}^{(k)} - \hat{\mathbf{x}}^{(j)} \right]_i \right|$$

subject to

$$\sum_{k=1}^{s} \lambda_{jk} = 1, \quad \text{and} \quad \lambda_{jk} \geq 0, \quad \forall k.$$

The constraints in the optimization problem guarantee the existence of the stationary joint distribution vector $\mathbf{x}$. Again we see that the above optimization problem formulates a linear programming problem for each j [22]:

$$\min_{\lambda} \ w_j$$

subject to

$$\begin{bmatrix} w_j \\ w_j \\ \vdots \\ w_j \end{bmatrix} \geq \hat{\mathbf{x}}^{(j)} - \mathcal{Q} \begin{bmatrix} \lambda_{j1} \\ \lambda_{j2} \\ \vdots \\ \lambda_{js} \end{bmatrix},$$

$$\begin{bmatrix} w_j \\ w_j \\ \vdots \\ w_j \end{bmatrix} \geq -\hat{\mathbf{x}}^{(j)} + \mathcal{Q} \begin{bmatrix} \lambda_{j1} \\ \lambda_{j2} \\ \vdots \\ \lambda_{js} \end{bmatrix},$$

for all $1 \le j \le s$,

$$w_j \ge 0, \quad \sum_{k=1}^{s} \lambda_{jk} = 1, \quad \text{and} \quad \lambda_{jk} \ge 0, \quad \forall k.$$

Here

$$\mathcal{Q} = [\hat{P}^{(j1)}\hat{\mathbf{x}}^{(1)} \mid \hat{P}^{(j2)}\hat{\mathbf{x}}^{(2)} \mid \ldots \mid \hat{P}^{(js)}\hat{\mathbf{x}}^{(s)}].$$

3.3.2 A numerical demonstration

We present a numerical example [7] of two sequences $\{X^{(1)}\}$ and $\{X^{(2)}\}$ of four states ($m = 4$):

$$S_1 = \{4, 3, 1, 3, 4, 4, 3, 3, 1, 2, 3, 4\} \quad \text{and} \quad S_2 = \{1, 2, 3, 4, 1, 4, 4, 3, 3, 1, 3, 1\}.$$

We then count the transition frequencies in

$$S_1: \; 4 \to 3 \to 1 \to 3 \to 4 \to 4 \to 3 \to 3 \to 1 \to 2 \to 3 \to 4$$

and

$$S_2: \; 1 \to 2 \to 3 \to 4 \to 1 \to 4 \to 4 \to 3 \to 3 \to 1 \to 3 \to 1$$

and we have

$$F^{(11)} = \begin{bmatrix} 0 & 0 & 2 & 0 \\ 1 & 0 & 0 & 0 \\ 1 & 1 & 1 & 2 \\ 0 & 0 & 2 & 1 \end{bmatrix} \quad \text{and} \quad F^{(22)} = \begin{bmatrix} 0 & 0 & 2 & 1 \\ 1 & 0 & 0 & 0 \\ 1 & 1 & 1 & 1 \\ 1 & 0 & 1 & 1 \end{bmatrix}.$$

We further count the inter-transition frequencies in

$$\begin{array}{llllllllllllllllllllllll}
S_1: 4 & & 3 & & 1 & & 3 & & 4 & & 4 & & 3 & & 3 & & 1 & & 2 & & 3 & & 4 \\
 & \nearrow & & \nearrow & & \nearrow & & \nearrow & & \nearrow & & \nearrow & & \nearrow & & \nearrow & & \nearrow & & \nearrow & & \nearrow & \\
S_2: 1 & & 2 & & 3 & & 4 & & 1 & & 4 & & 4 & & 3 & & 3 & & 1 & & 3 & & 1
\end{array}$$

and

$$\begin{array}{llllllllllllllllllllllll}
S_1: 4 & & 3 & & 1 & & 3 & & 4 & & 4 & & 3 & & 3 & & 1 & & 2 & & 3 & & 4 \\
 & \searrow & & \searrow & & \searrow & & \searrow & & \searrow & & \searrow & & \searrow & & \searrow & & \searrow & & \searrow & & \searrow & \\
S_2: 1 & & 2 & & 3 & & 4 & & 1 & & 4 & & 4 & & 3 & & 3 & & 1 & & 3 & & 1
\end{array}$$

and we have

$$F^{(21)} = \begin{bmatrix} 1 & 0 & 2 & 0 \\ 0 & 0 & 0 & 1 \\ 0 & 1 & 3 & 0 \\ 1 & 0 & 0 & 2 \end{bmatrix} \quad \text{and} \quad F^{(12)} = \begin{bmatrix} 0 & 1 & 1 & 0 \\ 0 & 0 & 1 & 0 \\ 2 & 0 & 1 & 2 \\ 1 & 0 & 1 & 1 \end{bmatrix}.$$

After making a normalization, we have the transition probability matrices:

$$\hat{P}^{(11)} = \begin{bmatrix} 0 & 0 & \frac{2}{5} & 0 \\ \frac{1}{2} & 0 & 0 & 0 \\ \frac{1}{2} & 1 & \frac{1}{5} & \frac{2}{3} \\ 0 & 0 & \frac{2}{5} & \frac{1}{3} \end{bmatrix}, \qquad \hat{P}^{(12)} = \begin{bmatrix} 0 & 1 & \frac{1}{4} & 0 \\ 0 & 0 & \frac{1}{4} & 0 \\ \frac{2}{3} & 0 & \frac{1}{4} & \frac{2}{3} \\ \frac{1}{3} & 0 & \frac{1}{4} & \frac{1}{3} \end{bmatrix},$$

$$\hat{P}^{(21)} = \begin{bmatrix} \frac{1}{2} & 0 & \frac{2}{5} & 0 \\ 0 & 0 & 0 & \frac{1}{3} \\ 0 & 1 & \frac{3}{5} & 0 \\ \frac{1}{2} & 0 & 0 & \frac{2}{3} \end{bmatrix}, \quad \hat{P}^{(22)} = \begin{bmatrix} 0 & 0 & \frac{1}{2} & \frac{1}{3} \\ \frac{1}{3} & 0 & 0 & 0 \\ \frac{1}{3} & 1 & \frac{1}{4} & \frac{1}{3} \\ \frac{1}{3} & 0 & \frac{1}{4} & \frac{1}{3} \end{bmatrix}.$$

We also have

$$\hat{\mathbf{x}}_1 = \left(\frac{1}{6}, \frac{1}{12}, \frac{5}{12}, \frac{1}{3}\right)^T \quad \text{and} \quad \hat{\mathbf{x}}_2 = \left(\frac{1}{3}, \frac{1}{12}, \frac{1}{3}, \frac{1}{4}\right)^T.$$

By solving the corresponding linear programming problems, the multivariate Markov chain model of the two categorical data sequences S_1 and S_2 is then given by

$$\begin{cases} \mathbf{x}_{t+1}^{(1)} = 0.5000\hat{P}^{(11)}\mathbf{x}_t^{(1)} + 0.5000\hat{P}^{(12)}\mathbf{x}_t^{(2)}, \\ \mathbf{x}_{t+1}^{(2)} = 0.8858\hat{P}^{(21)}\mathbf{x}_t^{(1)} + 0.1142\hat{P}^{(22)}\mathbf{x}_t^{(2)}. \end{cases}$$

3.3.3 Some practical applications

We present some applications of the multivariate Markov chain model. The applications include: (i) sales demand predictions, (ii) construction of genetic networks, and (iii) credit risk measurement.

(i) Sales demand predictions

We apply the multivariate Markov chain model to the sales demand data sequence of the inventory control problem (iii) in section 3.2.2 again. Here we have five related products ($s = 5$) and there are six possible states ($m = 6$) due to different sales volume. A product has six possible sales volumes: very fast-moving (very high sales volume), fast-moving, standard, slow-moving, very slow-moving (very low sales volume), and no sales volume [7]. The soft-drink company would like to predict sales demand for its big customers in order to minimize its inventory build-up and to maximize the demand satisfaction for this customer. We expect that sales demand sequences generated by the same customer are correlated to each other. Therefore, by exploring these relationships, we develop the multivariate Markov chain model for such sales demand sequences and obtain better prediction rules. We use the multivariate Markov chain model to predict the next state $\hat{\mathbf{x}}_t$ at time t by taking the state with the largest probability, i.e.,

$$\hat{\mathbf{x}}_t = j, \quad \text{if} \quad [\hat{\mathbf{x}}_t]_i \le [\hat{\mathbf{x}}_t]_j, \quad \text{for} \quad 1 \le i \le m.$$

To evaluate the performance and effectiveness of our multivariate Markov chain model, a prediction result is measured by the prediction accuracy r defined as

$$r = \frac{1}{T} \times \sum_{t=n+1}^{T} \delta_t \times 100\%,$$

where

$$\delta_t = \begin{cases} 1, & \text{if } \hat{\mathbf{x}}_t = \mathbf{x}_t, \\ 0, & \text{otherwise.} \end{cases}$$

To have a complete comparison, we also give the results for the first-order Markov chain model of individual sales demand sequence. The results are reported in Table 3.3 (c.f. [7]). There is a significant improvement in the prediction accuracy in Product A, while improvements are also observed in Products D and E. The results show the effectiveness of the multivariate Markov chain model.

Table 3.3: Prediction Accuracy of the Sales Demand Data.

	Product A	Product B	Product C	Product D	Product E
First-order Markov Model	46%	45%	63%	51%	53%
Multivariate Markov Model	50%	45%	63%	52%	55%

(ii) Construction of genetic networks

In genomic research, an important focus is to understand the mechanism in which cells execute and control the huge number of operations for normal functions, and also the way in which the cellular systems fail in disease. One of the possible reasons is the cell cycle. Cells in the body alternately divide in mitosis and inter-phase, and this sequence of activities exhibited by cells is called the cell cycle. Probabilistic Boolean Network (PBN), introduced by Kauffman [25], is a popular model for modeling genetic regulatory networks because it can help in discovering the underlying gene regulatory mechanisms. In a PBN, each gene is regarded as a vertex of the network and is only quantized into two levels (over-express (1) or under-express (0)), i.e., two states ($m = 2$). The dynamics of the PBNs can be studied in the context of standard Markov chain. However, the number of parameters in a PBN grows exponentially with respect to the number of genes expression sequences s. In [11], the multivariate Markov chain model is applied to gene expression data sequences of yeast. The number of parameters is $O(s^2)$, where s is the number of gene expression sequences for building the underlying genetic regulatory network. Interesting results about the cell cycles are also obtained.

(iii) Credit risk measurement

Credit risk modeling is a hot and important topic in risk management. In fact, there has been much interest in modeling the dependency of credit risks of securities in portfolios. There are two major approaches to model the dependency of credit risky securities, namely copulas [27] and Monte-Carlo simulation. The main advantages of copulas are that they are easy to apply and can capture the dependency of credit risks, while the Monte Carlo simulation technique is usually adopted together with copulas for modeling the dependency of credit risks [4].

Modeling the dynamics of transitions between credit ratings is vital for credit risk analysis in the finance industry. Siu et al. [33] apply the discrete-time homogeneous multivariate Markov chain model for dependent credit risks. Here the categorical data is the credit rating. The conditional probability distribution of the rating of a particular credit risk in the next period given the currently available information depends not only on its own current rating, but also on the current ratings of all other credit risks in the portfolio. The multivariate Markov chain model can handle both temporal (intra-) and cross-sectional (inter) dependencies of categorical data series for credit ratings (AAA, AA, A, BBB, BB, B, CCC, D). Their model actually can provide a consistent and convenient way to incorporate both historical rating data and another source of information, for instance, an expert opinion or a subjective view.

3.4 Extension of the high-dimensional Markov chain models

In this section, we discuss different extensions of the high-dimensional Markov chain models discussed in sections 3.2 and 3.3. The extensions include high-order multivariate Markov chain models, the simplified multivariate Markov chain models, and multivariate Markov chain models with negative associations.

3.4.1 High-order multivariate Markov chain models

Ching et al. [15] extend the models discussed in sections 3.2 and 3.3 to a high-order multivariate Markov chain model for multiple categorical data sequences. We note that in a conventional nth-order Markov chain model for s categorical data sequences of m states, there are $O(m^{ns})$ possible states. Thus the number of model parameters increases exponentially with respect to the number of categorical sequences and the order of the Markov chain model. The model proposed in [15] can capture both higher-order dependent intra-transition probabilities and inter-transition probabilities among the data sequences and the number of parameters is only $O(ns^2m^2)$.

In the model, it is assumed that there are s categorical sequences and each has m possible states in $\mathcal{M}$. We then consider an nth-order model. In the model, it is assumed that the state probability distribution of the jth sequence at time $t = r+1$ depends on the state probability distributions of all the sequences (including itself) at times

$$t = r, r-1, \ldots, r-n+1.$$

Using the same notations as in the previous two subsections, our proposed higher-order (nth-order) multivariate Markov chain model takes the following form:

$$\mathbf{x}_{r+1}^{(j)} = \sum_{k=1}^{s}\sum_{h=1}^{n} \lambda_{jk}^{(h)} P_h^{(jk)} \mathbf{x}_{r-h+1}^{(k)}, \quad j = 1, 2, \ldots, s, \quad r = n-1, n, \ldots,$$

with initials $\mathbf{x}_0^{(k)}, \mathbf{x}_1^{(k)}, \ldots, \mathbf{x}_{n-1}^{(k)}$, for $k = 1, 2, \ldots, s$. Here

$$\lambda_{jk}^{(h)} \geq 0, \quad 1 \leq j, k \leq s, \quad 1 \leq h \leq n; \quad \text{and} \quad \sum_{k=1}^{s}\sum_{h=1}^{n} \lambda_{jk}^{(h)} = 1, \quad j = 1, 2, \ldots, s.$$

The probability distribution of the jth sequence $\mathbf{x}_{r+1}^{(j)}$, at time $t = r+1$, depends on the weighted average of $P_h^{(jk)}\mathbf{x}_{r-h+1}^{(k)}$, where $P_h^{(jk)}$ is the hth step transition probability matrix which describes the hth step transition from the states in the kth sequence at time

$$t = r - h + 1$$

to the states in the jth sequence at time $t = r+1$ and $\lambda_{jk}^{(h)}$ is the weighting of this term.

They also show that a stationary vector of probability distributions of the resulting nth-order multivariate Markov chain exists. Parameter estimation methods based on a linear programming similar to those in sections 3.2 and 3.3 have also been developed. The 8th-order Markov chain model is then applied to the sales demand prediction problem in a soft-drink company in Hong Kong. The prediction results are then given in Table 3.4.

Table 3.4: Prediction Accuracy of the Sales Demand Data.

	Product A	Product B	Product C	Product D	Product E
8th-order Markov chain model	46%	48%	62%	52%	49%
First-order multivariate Markov chain model	50%	45%	63%	52%	55%
8th-order multivariate Markov chain model	50%	48%	63%	54%	55%

3.4.2 Simplified multivariate Markov chain models

All the models discussed so far assume that there are sufficient data sequences for estimation of model parameters. Otherwise, we have two problems. First, the estimation of the transition probability matrices may have large error. Second, the steady-state may not be reached. Ching et al. [21] consider the case when the length of the observed data sequences is very short. For the first problem, they propose to replace the transition probability matrix $P^{(ij)}$ in the model by the identity matrix and $P^{(ii)}$ by the following rank-one matrix

$$\hat{\mathbf{x}}^{(i)}\mathbf{1}^T, \tag{3.10}$$

where $\hat{\mathbf{x}}^{(i)}$ is the estimated stationary distribution of sequence i. For the second problem, the weights λ_{ij} should be chosen such that the multivariate Markov process converges very fast to the stationary distribution. The convergence rate of the process depends on the second largest eigenvalue in modulus of the matrix Q in (3.9). They then modified the simplified model for very short sequences by adding the extra constraints

$$0 \leq \lambda_{ii} \leq \beta < 1. \tag{3.11}$$

They serve the purpose of controlling the second largest eigenvalue of Q and hence the convergence rate of the multivariate Markov chain.

For the simplified model with the assumptions (3.10) and (3.11), Ching et al. further assume that

$$P^{(ii)} = \hat{\mathbf{x}}\mathbf{1}^T \quad \text{for all } i = 1, 2, \ldots, s.$$

In this case, for $\lambda_{ij} > 0$, the steady-state probability distribution $\hat{\mathbf{x}}$ is an invariant. The problem here is how to assign λ_{ij} such that the second largest eigenvalue of Q is small. They assume

$$[\Lambda]_{i,j} = \begin{cases} \alpha, & \text{if} \quad i = j, \\ \frac{1-\alpha}{m-1}, & \text{if} \quad i \neq j, \end{cases}$$

where $0 < \alpha < 1$. With these assumptions, the model takes the tensor product form:

$$Q = I \otimes \alpha P + (\Lambda - \alpha I) \otimes I.$$

The second largest eigenvalue of Q can be minimized by solving the following max-min problem:

$$\min_{0<\alpha<1} \left\{ \max \left\{ 1-\alpha, \frac{\alpha-1}{m-1}, \frac{\alpha m - 1}{m-1} \right\} \right\}.$$

It can be shown that the optimal value is $\alpha^* = \frac{m}{2m-1}$ and the optimal second largest eigenvalue, in this case, is $\frac{m}{2m-1}$. Numerical demonstrations in [21] indicate that the simplified multivariate Markov chain model has better Bayesian Information Criterion (BIC) than the original multivariate Markov chain model. Here BIC is the log-likelihood of the model, and it is a measure for the fitness of the model to the data.

3.4.3 Multivariate Markov chain models with negative associations

All the high-dimensional Markov chain models assume nonnegative weighting $\lambda_{i,j}$ and nonnegative matrix theory can be applied to analyze the resulting models. However, in some situation, the association can be negative. In [13], the following multivariate Markov chain model is proposed for this situation:

$$\begin{bmatrix} \mathbf{x}_{t+1}^{(1)} \\ \mathbf{x}_{t+1}^{(2)} \\ \vdots \\ \mathbf{x}_{t+1}^{(s)} \end{bmatrix} = \underbrace{\Lambda^{+} \begin{bmatrix} \mathbf{x}_{t}^{(1)} \\ \mathbf{x}_{t}^{(2)} \\ \vdots \\ \mathbf{x}_{t}^{(s)} \end{bmatrix}}_{\textit{Positive associated part}} + \underbrace{\frac{1}{m-1} \Lambda^{-} \begin{bmatrix} \mathbf{1}-\mathbf{x}_{t}^{(1)} \\ \mathbf{1}-\mathbf{x}_{t}^{(2)} \\ \vdots \\ \mathbf{1}-\mathbf{x}_{t}^{(s)} \end{bmatrix}}_{\textit{Negative associated part}}, \qquad (3.12)$$

where

$$\Lambda^{+} = \begin{bmatrix} \lambda_{1,1}P^{(11)} & \lambda_{1,2}I & \cdots & \lambda_{1,s}I \\ \lambda_{2,1}I & \lambda_{2,2}P^{(22)} & \cdots & \lambda_{2,s}I \\ \vdots & \vdots & \vdots & \vdots \\ \lambda_{s,1}I & \cdots & \lambda_{s,s-1}I & \lambda_{s,s}P^{(ss)} \end{bmatrix}$$

and

$$\Lambda^{-} = \begin{bmatrix} \lambda_{1,-1}P^{(11)} & \lambda_{1,-2}I & \cdots & \lambda_{1,-s}I \\ \lambda_{2,-1}I & \lambda_{2,-2}P^{(22)} & \cdots & \lambda_{2,-s}I \\ \vdots & \vdots & \vdots & \vdots \\ \lambda_{s,-1}I & \cdots & \lambda_{s,-s+1}I & \lambda_{s,-s}P^{(ss)} \end{bmatrix}.$$

Here $\lambda_{i,j} \geq 0$, for $i = 1, 2, \ldots, s$ and $j = \pm 1, \ldots, \pm s$, and

$$\sum_{j=-s}^{s} \lambda_{i,j} = 1 \, .$$

In (3.12), Λ^{+} and Λ^{-} represent the transition probability matrices for the positive and negative associations respectively. From the second term on the right-hand-side of (3.12), an increase (a decrease) in $\mathbf{x}_t$ leads to a decrease (or an increase) in $\mathbf{x}_{t+1}$. In other words, there is a negative association between the two probability vectors $\mathbf{x}_t$ and $\mathbf{x}_{t+1}$. With the normalization constant $\frac{1}{m-1}$, the vector $\mathbf{x}_{t+1}^{(j)}$ on the left-hand-side of (3.12) is a probability vector, for each $j = 1, 2, \ldots, s$. The model is then applied to credit rating problems in [18]. The number of parameters involved is only $O(sm^2 + s^2)$, where s is the number of credit entities in a portfolio and m is the number of rating categories. Using this new model, Ching et al. [18] then investigate empirically the effect of incorporating both positive and negative associations on portfolio credit risks by estimating the risk measures such as Value-at-Risk and Expected Shortfall.

3.5 Concluding remarks

In this chapter, we present high-dimensional Markov chain models for categorical data sequences with practical applications. Efficient numerical methods (e.g., linear programming) for solving the model parameters are also discussed. The followings are some further research issues. (i) While for the multivariate Markov chain models, they can be applied to construct genetic networks from gene expressions data sequences. An important focus of genomic research concerns understanding the mechanism in which cells execute and control the huge number of operations for normal functions, and also the way in which the cellular systems fail in disease. Future research can be done on designing effective control strategy, based on the constructed network, to avoid undesirable network states from happening [37]. (ii) Another possible application of the high-dimensional models is to extend the models to hidden Markov models. This is particularly important in financial modeling [16, 18, 19, 32]. (iii) Finally it will be useful to design some other new estimation methods for the model parameters [38] and apply the new models and methods to new practical problems.

References for Chapter 3

[1] P. Avery, *The analysis of intron data and their use in the detection of short signals*, J. Mol. Evoln., 26 (1987), pp. 335–340.

[2] A. Berman and R. Plemmons, *Nonnegative Matrices in the Mathematical Sciences*, Academic Press, New York, 1979.

[3] R. Chan and W. Ching, *Toeplitz-circulant preconditioners for Toeplitz systems and their applications to queueing networks with batch arrivals*, SIAM Journal of Scientific Computing, 17 (1996), pp. 762–772.

[4] U. Cherubini, E. Luciano, and W. Vecchiato, *Copulas Methods in Finance*, John Wiley & Sons, New York, 2004.

[5] W. Ching, *Iterative Methods for Queuing and Manufacturing Systems*, Springer Monograph in Mathematics, Springer, London, 2001.

[6] W. Ching, R. Chan, and X. Zhou, *Circulant preconditioners for Markov modulated Poisson processes and their applications to manufacturing systems*, SIAM Journal on Matrix Analysis and Its Applications, 18 (1997), pp. 464–481.

[7] W. Ching, E. Fung, and M. Ng, *A multivariate Markov chain model for categorical data sequences and its applications in demand predictions*, IMA Journal of Management Mathematics, 13 (2002), pp. 187–199.

[8] W. Ching, E. Fung, and M. Ng, *A higher-order Markov model for the newsboy's problem*, Journal of Operational Research Society, 54 (2003), pp. 291–298.

[9] W. Ching, E. Fung, and M. Ng, *Higher-order Markov chain models for categorical data sequences*, Nav. Res. Logist., 51 (2004), pp. 557–574.

[10] W. Ching, E. Fung, and M. Ng, *Building higher-order Markov chain models with EXCEL*, International Journal of Mathematical Education in Science and Technology, 35 (2004), pp. 921–932.

[11] W. Ching, E. Fung, M. Ng, and T. Akutsu, *On construction of stochastic genetic networks based on gene expression sequences*, International Journal of Neural Systems, 15 (2005), pp. 297–310.

[12] W. Ching, E. Fung, M. Ng, and T. Ng, *Multivariate Markov models for the correlation of multiple biological sequences*, International Workshop on Bioinformatics, PAKDD Seoul, Korea, 2003, pp. 23–34.

[13] W. Ching, L. Li, T. Li, and S. Zhang, *A new multivariate Markov chain model with applications to sales demand forecasting*, Proceedings of IESM 2007, in CD-ROM.

[14] W. Ching and M. Ng, *Markov Chains: Models, Algorithms and Applications*, International Series on Operations Research and Management Science, Springer, New York, 2006.

[15] W. Ching, M. Ng, and E. Fung, *Higher-order multivariate Markov chains and their applications*, Linear Algebra and Its Applications, 428 (2008), pp. 492–507.

[16] W. Ching, T. Siu, E. Fung, M. Ng, and W. Li, *Interactive hidden Markov models and their applications*, IMA Journal of Management Mathematics, 18 (2007), pp. 85–97.

[17] W. Ching, T. Siu, and L. Li, *Pricing exotic options under a higher-order hidden Markov model*, Applied Mathematics and Decision Science, Article ID 18014 (2007).

[18] W. Ching, T. Siu, L. Li, H. Jiang T. Li, and W. Li, *An improved parsimonious multivariate Markov chain model for credit risk*, Journal of Credit Risk, 5 (2009), pp. 1–25.

[19] W. Ching, T. Siu, L. Li, T. Li, and W. Li, *Modeling default data via an interactive hidden Markov model*, Computational Economics, 34 (2009), pp. 1–19.

[20] W. Ching, W. Yuen, and A. Loh, *An inventory model with returns and lateral transshipments*, Journal of Operational Research Society, 54 (2003), pp. 636–641.

[21] W. Ching, S. Zhang, and M. Ng, *On multi-dimensional Markov chain models*, Pacific Journal of Optimization, 3 (2007), pp. 235–243.

[22] V. Chvátal, *Linear Programming*, Freeman, New York, 1983.

[23] A. Hénaut and A. Danchin, *Analysis and predictions from* Escherichia coli *sequences, or* E. coli *in Silico*, Escherichia coli and Salmonella, Cellular and Molecular Biology, 114 (1996), pp. 2047–2065.

[24] T. Joachims, D. Freitag, and T. Mitchell, *WebWatch: A tour guide for the World Wide Web*, Proceedings of the Fifteenth International Joint Conference on Artificial Intelligence 1997, IJCAI 97, pp. 770–775.

[25] S. Kauffman, *The Origin of Orders*, Oxford University Press, New York, 1993.

[26] J. Kleffe and M. Borodovsky, *First and second moment of counts of words in random texts generated by Markov chains*, CABIO, 8 (1982), pp. 433–441.

[27] D. Li, *On default correlation: a copula function approach*, Journal of Fixed Income, 9 (2000), pp. 43–54.

[28] H. Lieberman, *Letizia: an agent that assists Web browsing*, Proceedings of the Fourteenth International Joint Conference on Artificial Intelligence 1995, IJCAI 95, pp. 924–929.

[29] I. MacDonald and W. Zucchini, *Hidden Markov and Other Models for Discrete-valued Time Series*, Chapman & Hall, London, 1997.

[30] A. Raftery, *A model for high-order Markov chains*, J. R. Statist. Soc. B, 47 (1985), pp. 528–539.

[31] A. Raftery and S. Tavare, *Estimation and modelling repeated patterns in high order Markov chains with the mixture transition distribution model*, Appl. Statist., 43 (1994), pp. 179–199.

[32] T. Siu, W. Ching, E. Fung, M. Ng, and X. Li, *A higher-order Markov-switching model for risk measurement*, Computers and Mathematics with Applications, 58 (2009), pp. 1–10.

[33] T. Siu, W. Ching, M. Ng, and E. Fung, *On a multivariate Markov chain model for credit risk measurement*, Quantitative Finance, 5 (2005), pp. 543–556.

[34] M. Waterman, *Introduction to Computational Biology*, Chapman & Hall, Cambridge, 1995.

[35] Q. Yang, Z. Huang, and M. Ng, *A data cube model for prediction-based Web prefetching*, Journal of Intelligent Information Systems, 20 (2003), pp. 11–30.

[36] W. Yuen, W. Ching, and M. Ng, *A direct method for solving block-Toeplitz with near-circulant-block systems with applications to hybrid manufacturing systems*, Numerical Linear Algebra with Applications, 12 (2005), pp. 957–966.

[37] S. Zhang, W. Ching, Y. Jiao, L. Wu, and R. Chan, *Construction and control of genetic regulatory networks: a multivariate Markov chain approach*, Journal of Biomedical Science and Engineering, 1 (2008), pp. 15–21.

[38] D. Zhu and W. Ching, *A new estimation method for multivariate Markov chain model with application in demand predictions*, to appear in the Proceedings of the 3rd International Conference on Business Intelligence and Financial Engineering (BIFE 2010), Hong Kong, August 13–15, 2010.

Chapter 4

An Additive Variational Model for Image Segmentation

Yan-nei Law [1] **Hwee-kuan Lee** [2]

Chao-qiang Liu [3] **Andy M. Yip** [4]

Abstract

We propose a variant of the Mumford-Shah model for the segmentation of overlapping semi-transparent objects with additive intensity value. Unlike the standard segmentation models, it does not only determine distinct objects in the image, but also recover possibly multiple membership of the pixels. To accomplish this goal, *a prior* knowledge about the smoothness of the objects is taken into account in the model. To solve the optimization problem involving geometric quantities efficiently, we apply a multi-phase level set method. Segmentation results on synthetic and real images validate the good performance of our model.

4.1 Introduction

We consider the problem of segmenting two overlapping semi-transparent objects whose intensity level in the intersection is approximately the sum of the level of the individual objects. Mathematically speaking, assume that O_1 and O_2 are two possibly overlapping semi-transparent objects in a domain Ω. We consider an image $u_0 : \Omega \rightarrow \mathbb{R}$ which can be

[1]Imaging Informatics Group, Bioinformatics Institute, A*Star, Singapore.

[2]Imaging Informatics Group, Bioinformatics Institute, A*Star, Singapore.

[3]Centre for Wavelets, Approximation and Information Processing, National University of Singapore, Singapore.

[4]Department of Mathematics, National University of Singapore, Singapore (email: andyyip@nus.edu.sg).

well modeled by a piecewise constant function $u : \Omega \to \mathbb{R}$ as follows:

$$u(x,y) = \begin{cases} c_1, & \text{if } (x,y) \in O_1 \setminus O_2, \\ c_2, & \text{if } (x,y) \in O_2 \setminus O_1, \\ c_1 + c_2, & \text{if } (x,y) \in O_1 \cap O_2, \\ c_3, & \text{if } (x,y) \in \Omega \setminus (O_1 \cup O_2). \end{cases}$$

We call the identification of O_1 and O_2 from a given image u_0 an *additive segmentation problem*. This is a fundamental image processing task with many real world applications, especially those involving measurement of concentration using imaging techniques. Examples include x-ray images, images of absorbent paper with mouse scent marks [7], and microscopy images recording protein expression [6]. Although many applications of such a model can be found, there has been little study of this problem.

Many types of segmentation have been studied. They include hard segmentation [5, 11], soft segmentation [4, 15], segmentation with depth [12, 13, 19], illusory object segmentation [18], and segmentation of touching objects [17]. Additive segmentation fundamentally differs from other segmentation problems in two aspects, i.e., multiple membership and additivity. Except for soft segmentation, all of the aforementioned segmentation problems do not allow multiple membership of the pixels. For soft segmentation, each pixel is given a degree of membership to various objects, but the additivity of intensity values is not considered.

Standard segmentation models aim at grouping pixels with similar characteristics together to determine distinct objects in the image. However, in additive segmentation, the characteristics (e.g., intensity levels) of the overlapped region are very different from those of the individual objects. Thus an additive segmentation model requires to group pixels with different characteristics together. As we will illustrate in Figure 4.1, this model requires us to consider not only the intensity value, but also the geometry of the objects. To accomplish this task, we assume certain smoothness of the objects. We found that Euler's elastica is particularly effective.

In this chapter, we propose a variational model for additive segmentation. We analyze a hard constrained and a soft constrained versions of the model respectively. We demonstrate that the soft one is more robust and practical. We also present a numerical scheme based on a multi-phase level set method. Segmentation results on synthetic and real images validate the good performance of our model.

4.2 Model

Our model is a variant of the piecewise constant Mumford-Shah segmentation model [11] which uses a least squares approach to fit the image with a piecewise constant function and uses a penalty method to regularize the geometry of the partition. However, instead of using length regularization as in the original Mumford-Shah model, we use Euler's elastica which is more effective in recovering multiple membership. We also propose a hard constrained and a soft constrained ways to model additivity.

4.2.1 Mumford-Shah segmentation model

The Mumford-Shah segmentation model [11] is one of the most popular segmentation models. It can handle gracefully complex situations and is very robust to noise. A distinctive feature is that it is region-based which allows it to segment objects without edges. For the same reason, it can group a cluster of smaller objects into a larger object.

For a given image u_0, the piecewise constant Mumford-Shah model seeks for a set of curves C and a set of constants $\mathbf{c} = \{c_1, c_2, \ldots, c_n\}$, which minimize the following energy functional:

$$F^{\text{MS}}(C, \mathbf{c}) = \sum_{i=1}^{n} \int_{\Omega_i} |u_0(x,y) - c_i|^2 dxdy + \alpha \times \text{Length}(C), \tag{4.1}$$

where $\alpha \geq 0$ is a parameter. The curves of C partition the image into n mutually exclusive segments Ω_i for $i = 1, 2, \ldots, n$. The idea is to partition the image such that the intensity of u_0 in each segment Ω_i is well-approximated by a constant c_i. The goodness-of-fit is measured by the fidelity term $\int_{\Omega_i} |u_0(x,y) - c_i|^2 dxdy$. A minimum description length principle is employed to regularize the geometry of the partition, which increases the robustness to noise and avoids spurious segments. The parameter α controls the trade-off between the goodness-of-fit and the length of the curves of C. It can be easily shown that for each fixed curve of C, the optimal constant c_i is given by the average of u_0 over Ω_i.

4.2.2 Euler's elastica

To motivate Euler's elastica, we consider the example shown in Figure 4.1. The original image u_0 consists of two crossing objects O_1 and O_2 with the same intensity value, and the intensity of their common part is the sum. Consider the four possible segmentations in Figure 4.1(b). All of them are consistent with the additive assumption. However, we would like to make *a prior* assumption that objects with smooth boundaries are preferred. In this case, Configuration 1 is preferred. To achieve this aim, we consider the notion of Euler's elastica [1, 2, 9, 10]. A curve Γ is said to be Euler's elastica if it is the equilibrium curve of the elasticity energy

$$E(\Gamma) = \int_{\Gamma} (\alpha + \beta|\kappa(s)|) ds,$$

where ds is the arc length element, $\kappa(s)$ is the scalar curvature of Γ, and both α and β are non-negative constants controlling the weighting of the length and the total absolute curvature terms. It can be seen that Configuration 1 has the minimal Euler's elastica energy. We note that the length regularization in the Mumford-Shah model corresponds to setting $\beta = 0$. In this case, the lengths of Configurations 1, 2, and 3 are identical such that Configuration 1 cannot be distinguished from the others. The use of total absolute curvature can resolve this problem. In our implementation, we use

$$E(\Gamma) = \int_{\Gamma} [\alpha + \beta\phi(\kappa)] ds,$$

where

$$\phi(x) = \begin{cases} x^2, & \text{if } |x| \leq 1, \\ |x|, & \text{if } |x| > 1. \end{cases}$$

This modification can preserve high curvature features such as corners better.

4.2.3 A hard additive model

We start off by stating some assumptions of the additive segmentation problem:

1. There is no self-overlapping, i.e., the overlapping region must be generated by two distinct objects.

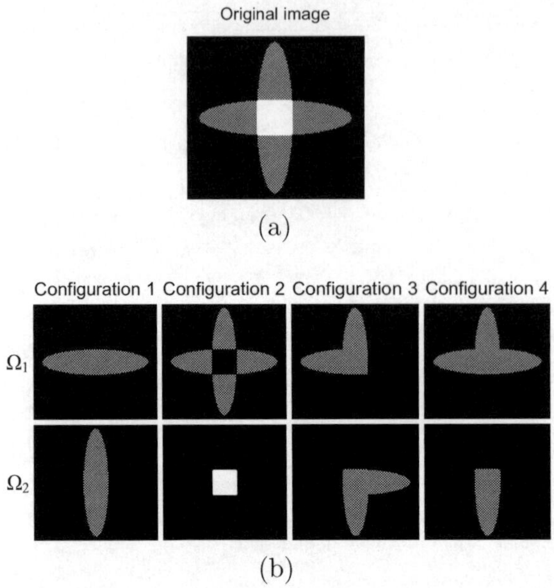

Figure 4.1: An example of two crossing objects. (a) Original image. (b) Four different segmentations which are consistent with the additivity assumption. Configuration 1 has the smallest Euler's elastica energy.

2. Intensity within each object can be well-approximated by a binary image.

3. Objects are preferred to have smooth boundary.

Given an image u_0, we seek for two regions Ω_1 and Ω_2, and a set of constants $\mathbf{c} = \{c_1, c_2, c_3, c_4\}$, which minimize the following energy:

$$\begin{aligned} F^{\text{hard}}(\Omega_1, \Omega_2, \mathbf{c}) &= \sum_{i=1}^{2} \int_{\partial\Omega_i \cap \Omega} [\alpha + \beta\phi(\kappa)] ds \\ &+ \int_{\Omega \setminus (\Omega_1 \cup \Omega_2)} [u_0(x,y) - c_3]^2 dxdy + \int_{\Omega_1 \setminus \Omega_2} [u_0(x,y) - c_1]^2 dxdy \\ &+ \int_{\Omega_2 \setminus \Omega_1} [u_0(x,y) - c_2]^2 dxdy + \int_{\Omega_1 \cap \Omega_2} [u_0(x,y) - c_4]^2 dxdy, \end{aligned}$$

subject to an additivity constraint

$$c_4 = c_1 + c_2.$$

This model enforces strict additivity in the overlapping region. We then refer it as a *hard additive model.* Although Ω_1 and Ω_2 generally do not constitute a partition of Ω, we still call the pair $\{\Omega_1, \Omega_2\}$ a segmentation of Ω for simplicity. It should be clear that the partition of Ω is given by $\{\Omega_1 \setminus \Omega_2, \Omega_2 \setminus \Omega_1, \Omega_1 \cap \Omega_2, \Omega \setminus (\Omega_1 \cup \Omega_2)\}$.

Given a fixed segmentation $\{\Omega_1, \Omega_2\}$, it can be easily shown that the optimal constants of $\mathbf{c}$ are given by the following formulas:

$$\begin{aligned}
\begin{bmatrix} c_1 \\ c_2 \end{bmatrix} &= \begin{bmatrix} |\Omega_1| & |\Omega_1 \cap \Omega_2| \\ |\Omega_1 \cap \Omega_2| & |\Omega_2| \end{bmatrix}^{-1} \begin{bmatrix} \int_{\Omega_1} u_0 dxdy \\ \int_{\Omega_2} u_0 dxdy \end{bmatrix}, \\
c_3 &= \frac{\int_{\Omega \setminus (\Omega_1 \cup \Omega_2)} u_0 dxdy}{|\Omega \setminus (\Omega_1 \cup \Omega_2)|}.
\end{aligned} \tag{4.2}$$

Here, $|\cdot|$ denotes the Lebesgue measure of its argument. The above linear system is non-singular if and only if neither of the objects is empty and $\Omega_1 \neq \Omega_2$.

In many applications, the additivity assumption may not be exactly satisfied. In this situation, due to the possible ill-conditioning of the above linear system, the resulting constants c_1 and c_2 may be totally different from the observed intensity values even for a small deviation in additivity in the overlapping region. According to some numerical tests of the hard additive model, we observe that the natural approach of level set method often converges to a bad local minimum; see section 4.4.1 for example. In order to solve this problem, we relax the hard constraint to a soft one.

4.2.4 A soft additive model

We consider the following relaxed model:

$$\begin{aligned}
F^{\mathrm{soft}}(\Omega_1, \Omega_2, \mathbf{c}) &= \sum_{i=1}^{2} \int_{\partial\Omega_i \cap \Omega} [\alpha + \beta\phi(\kappa)] ds \\
&+ \int_{\Omega \setminus (\Omega_1 \cup \Omega_2)} [u_0(x,y) - c_3]^2 dxdy + \int_{\Omega_1 \setminus \Omega_2} [u_0(x,y) - c_1]^2 dxdy \\
&+ \int_{\Omega_2 \setminus \Omega_1} [u_0(x,y) - c_2]^2 dxdy + \int_{\Omega_1 \cap \Omega_2} [u_0(x,y) - c_4]^2 dxdy \\
&+ \gamma(c_1 + c_2 - c_4)^2,
\end{aligned}$$

where $\gamma \geq 0$ is a constant controlling the additivity. In this model, the constraint $c_1 + c_2 = c_4$ is only loosely enforced. We call this model a *soft additive model.*

Note that when $\gamma = 0$, the soft model reduces to the piecewise constant Mumford-Shah segmentation model [11, 16]. When $\gamma \to \infty$, the solution of the soft model will approach to that of the hard model. Given a segmentation $\{\Omega_1, \Omega_2\}$, the optimal constants of $\mathbf{c}$ are obtained by the following explicit formulas:

$$\begin{bmatrix} c_1 \\ c_2 \\ c_4 \end{bmatrix} = \begin{bmatrix} |\Omega_1 \setminus \Omega_2| + \gamma & \gamma & -\gamma \\ \gamma & |\Omega_2 \setminus \Omega_1| + \gamma & -\gamma \\ -\gamma & -\gamma & |\Omega_1 \cap \Omega_2| + \gamma \end{bmatrix}^{-1} \begin{bmatrix} \int_{\Omega_1 \setminus \Omega_2} u_0 dxdy \\ \int_{\Omega_2 \setminus \Omega_1} u_0 dxdy \\ \int_{\Omega_1 \cap \Omega_2} u_0 dxdy \end{bmatrix}, \tag{4.3}$$

$$c_3 = \frac{\int_{\Omega \setminus (\Omega_1 \cup \Omega_2)} u_0 dxdy}{|\Omega \setminus (\Omega_1 \cup \Omega_2)|}.$$

When $\gamma > 0$, this linear system is non-singular if and only if neither of the objects is empty and $\Omega_1 \neq \Omega_2$.

4.3 Level set approach

It is not a trivial task to solve a shape optimization problem involving unknown topology and geometric quantities such as length and curvature. The level set method proposed by Osher and Sethian [14] is particularly well suited to handle topological changes and curvature dependent functionals, since the curvature of a contour has a very simple expression in terms of the level set function that represents the contour. It has been successfully applied for solving the Mumford-Shah problem [3, 16]. Using the multi-phase level set method [16], a set of curves C is represented by the zeroth level of a set of functions Ψ defined on Ω such that the functional $F^{\mathrm{MS}}(C, \mathbf{c})$ in (4.1) becomes $F^{\mathrm{MS}}(\Psi, \mathbf{c})$. The problem then becomes a standard variational problem which is much easier to handle. Besides the Mumford-Shah hard segmentation problem, level set methods have also been used by Zhu *et al.* in [18, 19] to compute illusory contours and minimize the Nitzberg-Mumford-Shiota functional respectively. In this section, we first reformulate the proposed soft additive model using a multi-phase level set representation of the objects Ω_1 and Ω_2. We then derive the gradient flow to minimize the functional.

4.3.1 Level set representation

To represent objects Ω_i, for $i = 1, 2$, we construct Lipschitz functions $\psi_i : \Omega \to \mathbb{R}$ which satisfy

$$\psi_i(x,y) \begin{cases} > 0, & \text{if } (x,y) \in \Omega_i, \\ = 0, & \text{if } (x,y) \in \partial\Omega_i, \\ < 0, & \text{if } (x,y) \in \Omega \setminus \overline{\Omega}_i. \end{cases}$$

Therefore, the proposed soft additive energy functional can be reformulated as

$$\begin{aligned} F^{\mathrm{soft}}(\psi_1, \psi_2, \mathbf{c}) &= \sum_{i=1}^{2} \int_\Omega [\alpha + \beta\phi(\kappa_i)] |\nabla \psi_i| \delta(\psi_i) dxdy \\ &+ \int_\Omega [u_0(x,y) - c_1]^2 H(\psi_1)[1 - H(\psi_2)] dxdy \\ &+ \int_\Omega [u_0(x,y) - c_2]^2 [1 - H(\psi_1)] H(\psi_2) dxdy \\ &+ \int_\Omega [u_0(x,y) - c_3]^2 [1 - H(\psi_1)][1 - H(\psi_2)] dxdy \\ &+ \int_\Omega [u_0(x,y) - c_4]^2 H(\psi_1) H(\psi_2) dxdy \\ &+ \gamma |c_1 + c_2 - c_4|^2, \end{aligned}$$

where $\kappa_i = \nabla \cdot (\frac{\nabla \psi_i}{|\nabla \psi_i|})$, δ is the Dirac Delta function, and $H : \mathbb{R} \to \mathbb{R}$ is the Heaviside function defined by

$$H(x) = \begin{cases} 1, & \text{if } x > 0, \\ 1/2, & \text{if } x = 0, \\ 0, & \text{if } x < 0. \end{cases}$$

4.3.2 Level set based gradient descent method

To optimize the functional $F^{\mathrm{soft}}(\Psi, \mathbf{c})$ with $\Psi = \{\psi_1, \psi_2\}$, we use an alternating minimization approach where the objective is minimized with respect to Ψ and $\mathbf{c}$ alternatively. To minimize F_{soft} with respect to Ψ, we use the method of steepest descent where an artificial time variable is introduced into each ψ_i, i.e., $\psi_i : \Omega \times [0, +\infty) \to \mathbb{R}$ and the steady state of the gradient flow is sought. Using standard variational calculus, the gradient flow for $F^{\text{soft}}(\cdot, \mathbf{c})$ can be derived as

$$\begin{aligned}\frac{\partial \psi_i}{\partial t} &= |\nabla \psi_i| \nabla \cdot \left\{ \frac{\nabla \psi_i}{|\nabla \psi_i|} [\alpha + \beta \phi(\kappa_i)] - \frac{1}{|\nabla \psi_i|} (I - P_{\mathbf{n}_i}) \nabla \left[\beta \phi'(\kappa_i) |\nabla \psi_i| \right] \right\} \\ &\quad - |\nabla \psi_i| [(u_0 - c_i)^2 - (u_0 - c_3)^2][1 - H(\psi_j)] \\ &\quad - |\nabla \psi_i| [(u_0 - c_4)^2 - (u_0 - c_j)^2] H(\psi_j), \qquad \text{for } i = 1, 2, \end{aligned} \tag{4.4}$$

where $j = 3 - i$, I is the identity operator, and $P_{\mathbf{n}_i} : \mathbb{R}^2 \to \mathbb{R}^2$, with $\mathbf{n}_i = \frac{\nabla \psi_i}{|\nabla \psi_i|}$, are projectors defined by $P_{\mathbf{n}_i}(\mathbf{v}) = (\mathbf{v} \cdot \mathbf{n}_i)\mathbf{n}_i$ for $\mathbf{v} \in \mathbb{R}^2$; see [19]. The boundary conditions are $\frac{\partial \psi_i}{\partial n} = 0$ on $\partial \Omega$. This system of equations is solved numerically by standard explicit finite difference methods. The detailed numerical implementation can be found in [19].

The gradient flow for $F^{\text{hard}}(\cdot, \mathbf{c})$ is almost the same. We only need to replace c_4 by $c_1 + c_2$ in (4.4). When we update the constants of $\mathbf{c}$, (4.2) is used instead of (4.3).

4.4 Numerical results

We validate our model based on: 1) several performance comparisons with some competitive methods; 2) the robustness to different parameters; 3) the applicability to diverse applications.

4.4.1 Comparison of various models

Hard model vs. soft model

In this test, we compare the hard and the soft models with $\alpha = \beta = 0$ using the image in Figure 4.2. The objective functional is optimized using the level set method introduced in section 4.3. The intensity values of two objects are $d_1 = 105$ and $d_2 = 130$ respectively. The areas (number of pixels) of various regions are $|O_1 \setminus O_2| = |O_2 \setminus O_1| = 2198$ and $|O_1 \cap O_2| = 611$.

The results of minimizing F^{hard} with the level set algorithm are shown in Figure 4.3, which is a wrong solution. However, the use of the soft model will lead to the correct segmentation; see Figure 4.4.

Mumford-Shah model vs. soft model

In this test, we use an image containing two irregular shapes with intensity values 70 and 180. The intensity value in their overlapping region is 250. The original image and the initial segmentation are shown in the first row of Figure 4.5. The results in the second row show that in this experiment, the proposed method is able to segment out the two objects successfully even if the initial condition is far away from the exact segmentation.

Next we compare the performance of the soft model with the four-phase Mumford-Shah segmentation model. The latter model corresponds to setting $\beta = \gamma = 0$ in the soft model. The result of the four-phase method is shown in the third row. As expected, the four-phase method alone is not sufficient to correctly segment images with overlapping semi-transparent

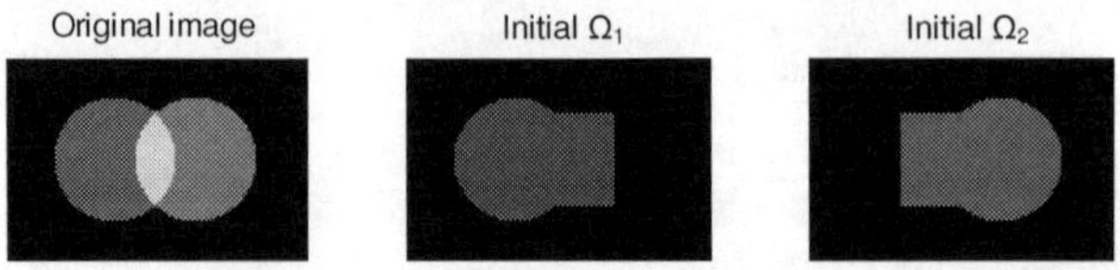

Figure 4.2: Original image and initial segmentation.

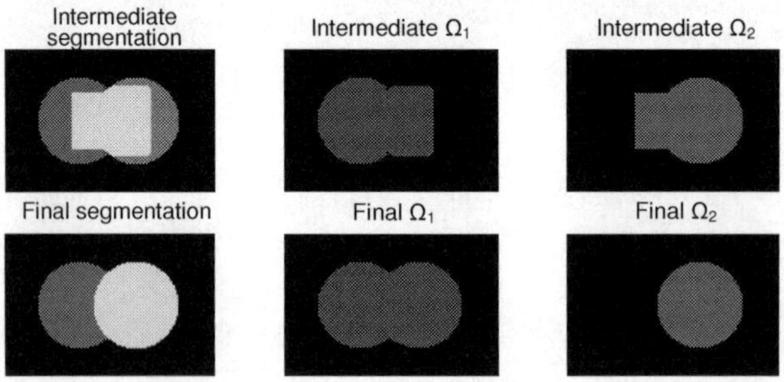

Figure 4.3: Intermediate and final segmentation results obtained by the hard additive model.

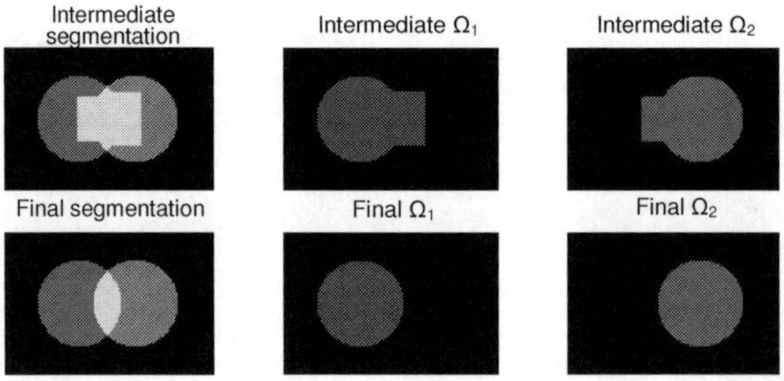

Figure 4.4: Intermediate and final segmentation results obtained by the soft additive model.

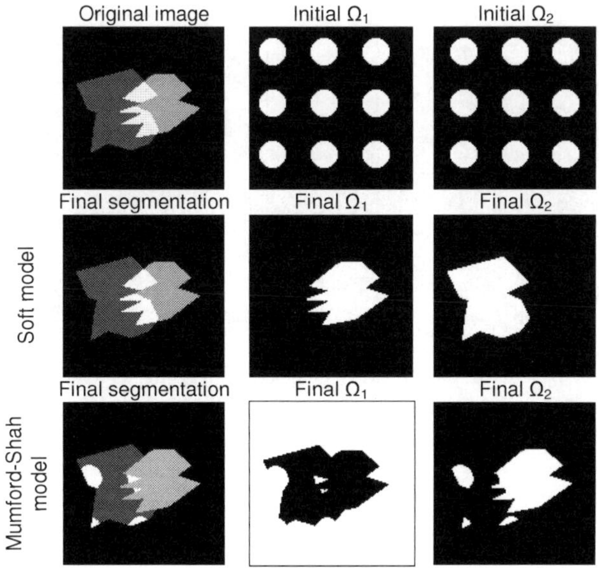

Figure 4.5: Results of the soft additive model and the four-phase Mumford-Shah model. First row: Original image with size 128×128 and the initial segmentation. Second row: Soft additive model with $\alpha = 1000$, $\beta = 0$, and $\gamma = 5000$. Third row: Four-phase Mumford-Shah model with $\alpha = 1000$ and $\beta = \gamma = 0$.

objects. Indeed, the iterates are trapped in a local minimum. For example, in the segment Ω_2, there are three small isolated pieces. Shrinking these pieces will reduce the total length of Ω_2 and keep the fitting term unchanged. Unfortunately, the total length of Ω_1 may increase such that the segmentation is reluctant to change. However, when the soft model is used, the fitting term may be reduced by shrinking the small spurious pieces.

4.4.2 Robustness to different parameters

In this experiment, we show that the soft model is robust to initial guess, deviation in additivity, and inhomogeneity of intensity.

Robustness to initial guess

The original image and the initial segmentation are shown in the first row of Figure 4.6. The intensity values of the two objects and the overlapping region are 70, 180, and 250 respectively. We use a 'seed' initial condition which is a commonly used generic condition. In the image, two objects cross over each other such that the non-overlapping region of each object becomes disconnected, which adds some difficulties to the identification of the correct segmentation since many connections are possible. Thus it invokes the use of some smooth connectivity assumptions. The results in the second row show that the proposed method is able to segment out the two objects successfully.

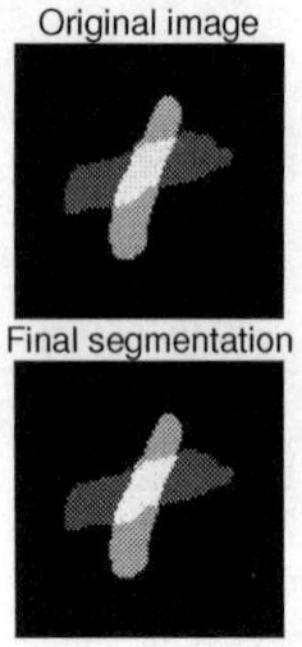

Figure 4.6: Segmentation results on the synthetic crossing marks image. First row: Original image with size 100×100 and the initial condition of two level set functions. Second row: Soft additive model with $\alpha = 100$, $\beta = 50$, and $\gamma = 1000$.

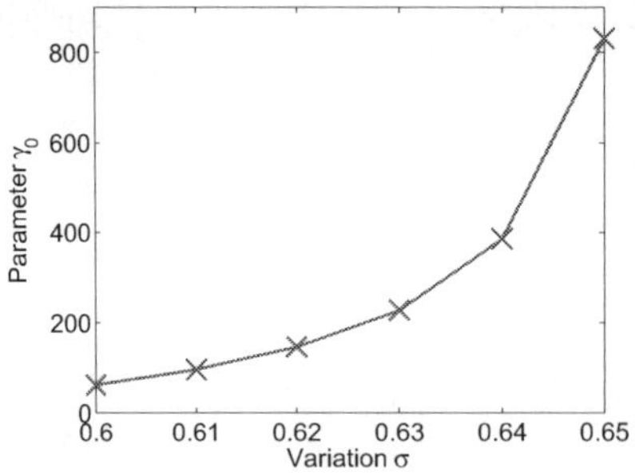

Figure 4.7: Cut-off values γ_0 determined empirically vs. the deviation σ. The empirical results are obtained using the level set approach with $\alpha = \beta = 0$.

Robustness to deviation in additivity

In this test, we use again the image in Figure 4.6 except the intensity values of the objects are now 100 and 130. The intensity value of the overlapping region is $\sigma(100 + 130)$ with a varying σ. We apply the hard model and the soft model with $\alpha = \beta = 0$ using level set implementations respectively. We use the ideal segmentation as the initial segmentation. We observe that the hard model leads to a wrong solution when $\sigma < 0.65$. For each $\sigma \in \{0.60, 0.61, 0.62, 0.63, 0.64, 0.65\}$, we minimize F^{soft} with different γ's using the level set approach and empirically determine the cut-off value γ_0 (shown in Figure 4.7) below which the method stabilizes at the ideal segmentation. This shows that the problem may be resolved by minimizing F^{soft} instead.

Robustness to inhomogeneity of intensity

In this test, we use a real x-ray image to demonstrate how our method works in real world scenarios. The original image and a manual segmentation are shown in the first row of Figure 4.8. The image exhibits a certain degree of additivity, but the intensity level within each region is highly inhomogeneous. The intensity distribution of each segment is summarized

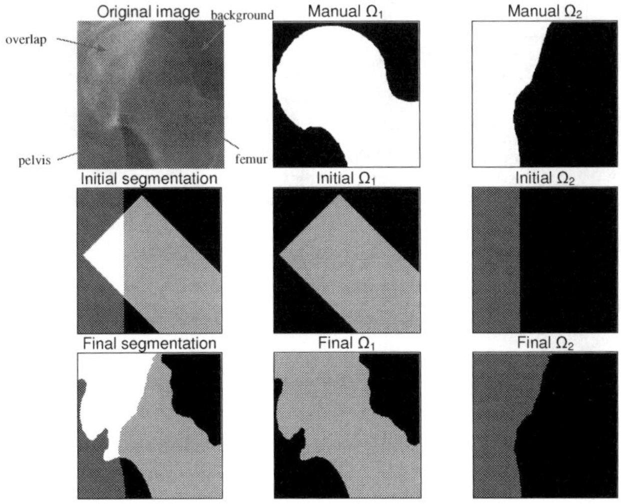

Figure 4.8: Segmentation results of the hip joint image. First row: Original image with size 240×240 and the initial segmentation. Second row: Results of the soft additive model with $\alpha = 2200$, $\beta = 10$, and $\gamma = 1000$.

below:

	mean	standard deviation
pelvis	60.6094	37.3692
femur	34.5987	15.9237
overlap	106.4818	22.4055
background	3.3792	7.6219

Note that the intensity of the overlapping region is almost the same as the sum of the intensities of the femur and pelvis. The second row of Figure 4.8 shows the initial segmentation being used. The results in the third row show that the proposed soft model is able to obtain a fairly good segmentation. The pelvis is captured correctly. The femur is slightly off. We note that the intensity of the pelvis near the top left corner is as bright as the overlapping part which makes the image difficult to segment correctly. As a result, the method assigns this corner as part of the overlapping region.

4.4.3 Other applications

To further illustrate the usefulness of the proposed model, we apply it to two other applications: density-based clustering and colored imaging.

Density-based clustering

In this test, we use a binary image to represent a data set consisting of two dimensional data points. The intensity of a pixel is 255 if it is populated with at least one point or 0 otherwise. The data set contains two overlapping clusters. We assume that each cluster is of approximately uniform density and the density is additive at where clusters overlap. Here, the density of a region R may be taken as $\frac{1}{|R|}\int_R u_0 dxdy$. The task is to find the clusters. The original image and the initial segmentation are shown in the first row of Figure 4.9.

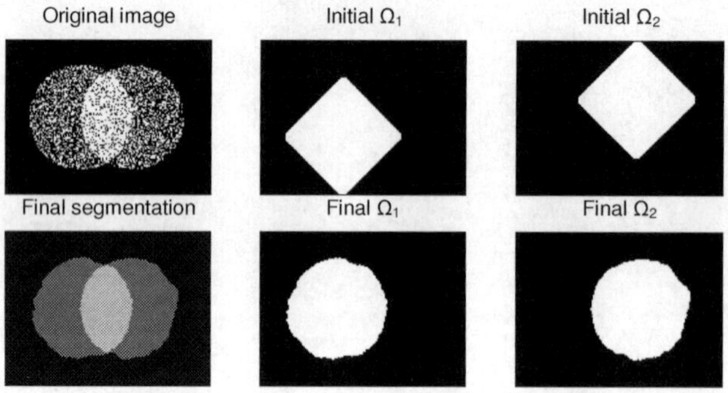

Figure 4.9: Segmentation results of the image representing a 2-D data set. First row: Original image with size 100×100 and the initial segmentation. Second row: Results of the soft additive model with $\alpha = 10000$, $\beta = 0$, and $\gamma = 10000$.

Note that this task is different from the aforementioned additive segmentation in which the intensity levels are no longer additive in the present case. It is the density which is additive. Indeed, the intensity levels within a cluster are far from uniform such that the fidelity term is not even close to 0. However, our method can segment the two clusters effectively. The results are shown in the second row of Figure 4.9.

Multi-channel images

In this test, we use a multi-channel microscopy image of a cell nucleus which captures the p53 and mdm2 protein concentration profiles in the green and red channels respectively. The task is to segment two colored overlapping semi-transparent objects. The aforementioned (single-channel) soft additive model can be easily generalized to handle multi-channel images.

In this case, one replaces the scalar function u_0 and scalar constants c_i in F^{soft} in section 4.2.4 with a vector function $\mathbf{u}_0$ and vector constants $\mathbf{c}_i$ for $i = 1, 2, 3, 4$. The integral $\int [u_0(x, y) - c_i]^2 dxdy$ is then replaced by $\int \|\mathbf{u}_0(x, y) - \mathbf{c}_i\|^2 dxdy$. The results are depicted in Figure 4.10. In the second row, the boundary of the objects is overlaid on the top of each channel. Since the signals from red and green channels are quite well-separated, the ground truth of each channel can be easily obtained for comparison. The results show that the soft model is able to separate the two proteins and the background accurately.

4.5 Discussion

In this chapter, we propose a variational model for the segmentation of overlapping semi-transparent objects with additive intensity value. Unlike the standard segmentation models, a successful additive segmentation model needs to recover possibly multiple membership of the pixels. We achieve this target by modeling additivity explicitly and incorporating geometric prior using Euler's elastica. We also show that the soft additive model is more robust than the hard model with respect to deviation of additivity and global convergence.

We remark that we use gradient descent and level set implementation to optimize the objective because of their simplicity. Our focus is the modeling aspect, leaving the computational procedure not optimized for speed. However, one can readily replace the optimization

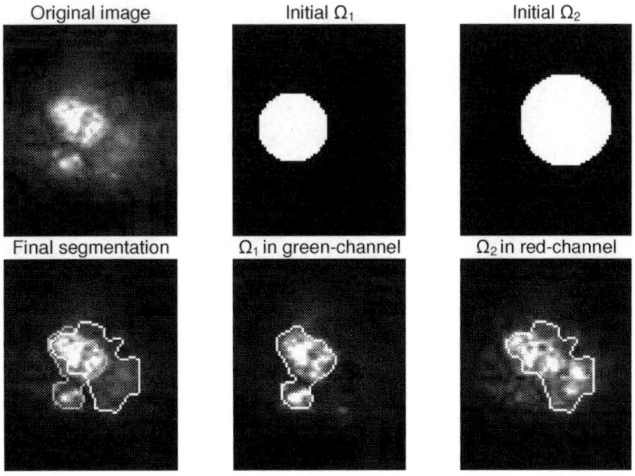

Figure 4.10: Segmentation results of the microscopy image. First row: Original image with size 92×73 and the initial segmentation. Second row: The boundary of the objects obtained by the soft additive model with $\alpha = 100$, $\beta = 0$, and $\gamma = 1000$, overlaid in the color image (left), green channel image (center), and red channel image (right).

procedure by any faster alternatives. Moreover, as with any local optimization method, the result depends very much on the initial condition. We alleviate this problem by choosing some good initial conditions. However, global optimization algorithms can be used [8].

As the first attempt to solve additive segmentation problems, we identified some fundamental difficulties even for the case of two objects. Future work should be considered for the segmentation of multiple objects and non-constant intensity objects.

Acknowledgement

Law and Lee are supported in part by the Biomedical Research Council of A*STAR, Singapore. Liu is supported by the Wavelets and Information Processing Programme under a grant from DSTA, Singapore. Yip is supported in part by Academic Research Grant R-146-000-116-112 from the National University of Singapore, Singapore.

References for Chapter 4

[1] T. Chan, S. H. Kang, and J. Shen, *Euler's elastica and curvature based inpaintings*, SIAM J. Appl. Math., 63 (2002), pp. 564–592.

[2] T. Chan, S. Esedoglu, and M. Nikolova, *Algorithms for finding global minimizers of denoising and segmentation models*, SIAM J. Appl. Math., 66 (2006), pp. 1632–1648.

[3] T. Chan and L. Vese, *Active contours without edges*, IEEE Trans. Image Process., 10 (2001), pp. 266–277.

[4] D. Cremers, M. Rousson, and R. Deriche, *A review of statistical approaches to level set segmentation: integrating color, texture, motion and shape*, Int. J. Comput. Vis., 72 (2007), pp. 195–215.

[5] J. Freixenet, X. Muñoz, D. Raba, J. Martí, and X. Cufí, *Yet another survey on image segmentation: region and boundary information integration*, Lecture Notes in Comp. Sci., Proc. of ECCV '02 (III), 2352 (2002), pp. 408–422.

[6] N. Geva-Zatorsky, N. Rosenfeld, S. Itzkovitz, R. Milo, E. Sigal, A. Dekel, T. Yarnitzky, P. Liron, Y. Polak, G. Lahav, and U. Alon, *Oscillations and variability in the p53 system*, Mol. Syst. Biol. 2: 2006.0033.

[7] J. L. Hurst, *Scent marking and social communication*, Animal Communication Networks, pp. 219–243, Cambridge University Press, Cambridge, 2005.

[8] Y. N. Law, H. K. Lee, and A. Yip, *A multi-resolution stochastic level set method for Mumford-Shah image segmentation*, IEEE Trans. Image Process., 17 (2008), pp. 2289–2300.

[9] A. Masnou and J. M. Morel, *Level-lines based disocclusion*, Proc. of 5th IEEE Int. Conf. on Image Process., 3 (1998), pp. 259–263.

[10] D. Mumford, *Elastica and computer vision*, Algebraic Geometry and Its Applications, pp. 491–506, Springer, New York, 1994.

[11] D. Mumford and J. Shah, *Optimal approximation by piecewise smooth functions and associated variational problems*, Comm. Pure Appl. Math., 42 (1989), pp. 577–685.

[12] M. Nitzberg and D. Mumford, *The 2.1d sketch*, Proc. of the 3rd Int. Conf. on Computer Vision, pp. 138–144, 1990.

[13] M. Nitzberg, D. Mumford, and T. Shiota, *Filtering, Segmentation, and Depth*, Springer-Verlag, New York, Inc., Secaucus, NJ, 1993.

[14] S. Osher and J. A. Sethian, *Fronts propogating with curvature-dependent speed: algorithms based on Hamilton-Jacobi formulation*, J. Comput. Phys., 79 (1988), pp. 12–49.

[15] J. Shen, *A stochastic-variational model for soft Mumford-Shah segmentation*, International Journal of Biomedical Imaging, 2006, Article ID: 92329.

[16] L. A. Vese and T. Chan, *A multiphase level set framework for image segmentation using the Mumford and Shah model*, Int. J. Comput. Vis., 50 (2002), pp. 271–293.

[17] P. Yan, X. Zhou, and S. Wong, *Automatic segmentation of high-throughput RNAi fluorescent cellular images*, IEEE Trans. Info. Tech. Biomed., 12 (2008), pp. 109–117.

[18] W. Zhu, *Variational Models for Illusory Contours and Shape Prior Segmentation*, PhD thesis, Dept. of Math., UCLA, 2005.

[19] W. Zhu, T. Chan, and S. Esedoḡlu, *Segmentation with depth: a level set approach*, SIAM J. Sci. Comput., 28 (2006), pp. 1957–1973.

Chapter 5

Total Variation Image Restoration with Automatic Selection of Regularization Parameters

Hai-yong Liao [1]   Michael K. Ng [2]

Abstract

Total variation based methods are popular for image restoration. One typical problem in these methods is how to select the regularization parameters. The common approach is try-and-error. In the literature, there are several regularization parameter selection methods for the Tikhonov regularization problems, for example, the discrepancy principle and the generalized cross-validation methods. However, these selection methods cannot be directly applied to the total variation regularization problems. In this chapter, we mainly study the regularization parameter problems in total variation based image deblurring problems. Within an alternating minimization framework, we exploit the generalized cross-validation technique to select the regularization parameter in each iteration. We also extend the idea to blind deconvolution problems. Experimental results are given to demonstrate the performance of the proposed methods.

5.1 Introduction

Image restoration is to restore or recover an ideal image from observed image via an inverse process. It has many important applications in the real world. Image deblurring is one of such cases. In photography, the blurring of images often results from the motion of objects,

[1]Center for Mathematical Imaging and Vision and Department of Mathematics, Hong Kong Baptist University, Hong Kong (email: 06459358@hkbu.edu.hk).

[2]Centre for Mathematical Imaging and Vision and Department of Mathematics, Hong Kong Baptist University, Hong Kong (email: mng@math.hkbu.edu.hk).

unfocused cameras, or calibration errors with imaging devices. Mathematically, the model of blurring process can be formulated as

$$g = h \otimes f + w, \tag{5.1}$$

where g is the observed image, h is the point spread function (PSF), i.e., the blurring kernel, f is the original image, and w is the Gaussian white noise. The notation $\otimes$ denotes the convolution operator. The goal for image deblurring is to restore f from the given observed image g and PSF h. Furthmore, if h is unknown, the problem would be more challenging and it is known as the blind deconvolution.

It is well known that the image deblurring problems are in general ill-posed inverse problems. So the restoration result is very sensitive to any perturbation. In order to obtain a meaningful solution, a regularization method is employed to achieve the stability of the approximation. In 1992, Rudin et al. [15] proposed the use of the total variation (TV)

$$TV(f) = \iint \sqrt{f_x^2 + f_y^2}\, dxdy$$

as a regularization functional, where $f(x, y)$ is the image. In the literature, there are several possible strategies for the *selection of regularization parameters*, e.g., the discrepancy principle method, the generalized cross-validation (GCV) method, and the L-curve method. However, these regularization parameter selection methods have not been developed for the TV image restoration problems, since the TV norm is nonlinear.

Recently, Lin and Wohlberg [11] extended the method of unbiased predictive risk estimator to the TV regularization problem by considering a quadratic approximation to the TV term. Nevertheless, they only reported the experimental results on a small problem. Babacan et al. [1] proposed a variational method for solving deconvolution problems with the TV regularization. Their idea is to use a stochastic method to approximate a posteriori distribution by a product of distributions using the Kullback-Leibler divergence. Their resulting method can update the regularization parameter in the TV image restoration in an iterative manner. Molina et al. [12] also studied this kind of variational method for blind deconvolution applications.

In [14], Reeves and Mersereau modeled the image as a two-dimensional autoregressive process, and the degraded image as a two-dimensional moving average process. They used the GCV method to estimate the parameters of these autoregressive and moving average processes. When there are several parameters to be estimated, the associated GCV functions are quite complicated. Reeves and Mersereau proposed to solve this problem by estimating a parameter one by one, and fixing the other parameters in the process. It is obvious that this procedure can be quite time-consuming when there are many parameters.

The outline of this chapter is as follows. In section 5.2, we consider the optimization models for image deconvolution problems, including non-blind and blind. We study the properties of optimization models and employ the alternating minimization scheme to solve the models. In this scheme, general cross validation can be used to select the regularization parameters. In section 5.3, numerical results are given to demonstrate the performance of the proposed methods. We compare the proposed methods with some state-of-the-art methods. Our experimental results show that the proposed methods are quite promising. Finally concluding remarks are given in section 5.4. The main contributions of this chapter are two effective algorithms for non-blind and blind image deconvolutions with automatic selection of regularization parameters.

5.2 Alternating minimization scheme with GCV for TV image restoration

In this section, we study two different methods for image deblurring problems. The non-blind deconvolution method is given in section 5.2.1, and the blind deconvolution method is given in section 5.2.2.

5.2.1 Non-blind image deconvolution

After discretization, the equation (5.1) can be rewritten in a matrix-vector form:

$$\mathbf{g} = H\mathbf{f} + \mathbf{w},$$

where the vectors $\mathbf{g}$, $\mathbf{f}$, $\mathbf{w}$, and the blurring matrix H correspond to the functions g, f, w, and the PSF h, respectively. When the PSF h is known, we mainly consider the following minimization model for the image deblurring problem:

$$\min_{\mathbf{f}} \frac{1}{\alpha}\|H\mathbf{f} - \mathbf{g}\|_2^2 + \|\mathbf{f}\|_{TV}, \tag{5.2}$$

where α is a positive regularization parameter which measures the trade off between a good fit and a TV regularization. The discrete version of the TV regularization term $\|\cdot\|_{TV}$ is given as follows. The discrete gradient operator $\nabla : \mathbb{R}^{n^2} \to \mathbb{R}^{n^2} \times \mathbb{R}^{n^2}$ is defined by

$$(\nabla\mathbf{f})_{j,k} = \left((\nabla\mathbf{f})^x_{j,k}, (\nabla\mathbf{f})^y_{j,k}\right) \in \mathbb{R}^2,$$

where

$$(\nabla\mathbf{f})^x_{j,k} = \begin{cases} \mathbf{f}_{j+1,k} - \mathbf{f}_{j,k}, & \text{if } j < n, \\ 0, & \text{if } j = n, \end{cases} \qquad (\nabla\mathbf{f})^y_{j,k} = \begin{cases} \mathbf{f}_{j,k+1} - \mathbf{f}_{j,k}, & \text{if } k < n, \\ 0, & \text{if } k = n, \end{cases}$$

for $j, k = 1, \ldots, n$, and $\mathbf{f}_{j,k}$ refers to the $(jn - n + k)$th entry of the vector $\mathbf{f}$, i.e., the (j,k)th pixel location of the image. The discrete TV of $\mathbf{f}$ is defined by

$$\|\mathbf{f}\|_{TV} \equiv \sum_{1\le j,k\le n} \|(\nabla\mathbf{f})_{j,k}\|_2 = \sum_{1\le j,k\le n} \sqrt{|(\nabla\mathbf{f})^x_{j,k}|^2 + |(\nabla\mathbf{f})^y_{j,k}|^2}.$$

We present a variable-splitting method and an alternating minimization scheme to solve (5.2) [9, 8]. Instead of (5.2), we minimize the following model:

$$\min_{\mathbf{f},\bar{\mathbf{f}}} \frac{1}{\alpha}\|H\mathbf{f} - \mathbf{g}\|_2^2 + \frac{\beta}{2}\|\bar{\mathbf{f}} - \mathbf{f}\|_2^2 + \|\bar{\mathbf{f}}\|_{TV}, \tag{5.3}$$

where $\beta > 0$ and $\bar{\mathbf{f}}$ is an auxiliary variable. More precisely, in (5.3), we replace $\mathbf{f}$ in the TV-norm term with $\bar{\mathbf{f}}$ and make $\bar{\mathbf{f}}$ close to $\mathbf{f}$ by an additional quadratic term. It is proved that when β is sufficiently large, the minimization problem in (5.3) is equivalent to the original problem (5.2) [5, 17].

An alternating minimization method is employed to solve (5.3) by iteratively minimizing the two sub-problems: one is solved for the determination of $\mathbf{f}$, and the other is solved for the determination of $\bar{\mathbf{f}}$. In the algorithm, we simultaneously increase the value of β so that the solution to the original problem (5.2) can be approximated by solving (5.3). The procedure is summarized in Algorithm 1.

Algorithm 1 Image deconvolution

1. Initialize β and θ (the multiplier), and set ϵ to be the threshold value for relative difference.

2. Fix $\bar{\mathbf{f}}$ and solve for $\mathbf{f}$,

$$\min_{\mathbf{f}} \|H\mathbf{f} - \mathbf{g}\|_2^2 + \gamma\|\mathbf{f} - \bar{\mathbf{f}}\|_2^2,$$

   where $\gamma = \frac{\alpha\beta}{2}$.

3. Fix $\mathbf{f}$ and solve for $\bar{\mathbf{f}}$,

$$\min_{\bar{\mathbf{f}}} \frac{\beta}{2}\|\bar{\mathbf{f}} - \mathbf{f}\|_2^2 + \|\bar{\mathbf{f}}\|_{TV}.$$

4. Increase β by $\beta = \theta \cdot \beta$.

5. **if** the relative difference between two successive iterate $\bar{\mathbf{f}}$ is less than ϵ, **then**

6. Goto step 10

7. **else**

8. Goto step 2

9. **end if**

10. Output: $\bar{\mathbf{f}}$.

The optimization problem in step 2 is equivalent to solving the following linear system:

$$(H^T H + \gamma I)\mathbf{f} = H^T\mathbf{g} + \gamma\bar{\mathbf{f}}, \tag{5.4}$$

where I is the identity matrix. This linear system can be solved by fast transform-based methods and preconditioning techniques for Toeplitz-like matrices [13]. For example, when the PSF is symmetric and the periodic boundary condition is used, the matrix H can be diagonalized by $H = F^*\Lambda F$, where F is the Fourier transform matrix, "*" denotes the conjugate transpose of a matrix, and Λ is a diagonal matrix containing the eigenvalues of H. After some simple substitutions, it is easy to obtain the solution for (5.4) in a close form

$$\mathbf{f} = F^* \left[(\Lambda^*\hat{\mathbf{g}} + \gamma\hat{\bar{\mathbf{f}}})./(\Lambda^*\Lambda + \gamma I) \right],$$

where $\hat{\mathbf{g}} = F\mathbf{g}$, $\hat{\bar{\mathbf{f}}} = F\bar{\mathbf{f}}$, and "./" denotes the entry-wise division operator. The TV optimization problem in step 3 can be solved by Chamblle's projection algorithm [3].

In Algorithm 1, we need to input the value of regularization parameter γ in step 2. We propose to employ the GCV method to determine a suitable value of the regularization parameter.

Estimation of regularization parameter

The GCV [7] is a technique that estimates a parameter directly without any knowledge on noises. Furthermore, it is essentially the leave-one-out cross-validation. The basic idea is to take one observation out of all observed data as the validation data, and then to use the remaining observations to predict the observation for validation. In each round, one observation in the observed data is used once as the validation data. If a parameter is a good choice, the overall prediction error would be minimized. Therefore, the best parameter can be determined by finding the minimizer of the overall prediction error, i.e., the GCV function.

Next we consider the regularization parameter γ for the restoration of images in step 2 of Algorithm 1. The linear least squares problem can be solved in the Fourier transform domain, which is equivalent to

$$\min_{\hat{\mathbf{f}}} \|\Lambda\hat{\mathbf{f}} - \hat{\mathbf{g}}\|_2^2 + \gamma\|\hat{\mathbf{f}} - \hat{\bar{\mathbf{f}}}\|_2^2, \tag{5.5}$$

where $\hat{\mathbf{f}} = F\mathbf{f}$. By applying the result in [7, 6], we can construct the GCV function of γ for (5.5):

$$GCV_{\mathbf{f}}(\gamma) = \frac{n^2 \sum_{i=1}^{n^2} \left(\frac{|(\hat{\mathbf{g}} - \Lambda\hat{\bar{\mathbf{f}}})_i|}{|[\Lambda]_{i,i}|^2 + \gamma} \right)^2}{\left(\sum_{i=1}^{n^2} \frac{1}{|[\Lambda]_{i,i}|^2 + \gamma} \right)^2}. \tag{5.6}$$

Numerical techniques can be applied to find the minimizer of $GCV_{\mathbf{f}}(\gamma)$. The search of γ in (5.6) involves scalar arithmetic operations and Λ is calculated by the fast Fourier transform. Therefore, (5.6) can be solved fast.

5.2.2 Extension to blind image deconvolution

The blind image deconvolution would be more challenging, as the PSF h is unknown. We adopt similar ideas in [2] and consider the following minimization model for the blind image deconvolution problem:

$$\min_{f,h} E(f,h) \equiv \min_{f,h} \|h \otimes f - g\|_{L^2}^2 + \alpha_1 TV(f) + \alpha_2 \|\Delta h\|_{L^2}^2, \tag{5.7}$$

where Δ is the Laplacian operator, α_1 and α_2 are positive regularization parameters which measure the trade-off between a good fit and regularization terms.

The minimization problem (5.7) is on a non-convex functional of two variables and admits infinitely many solutions. However, the functional in (5.7) is convex with respect to one variable f (or h) if the other variable h (or f) is fixed. Furthermore, when h is fixed, minimizing (5.7) leads to a more difficult problem, which is to minimize a convex and non-smooth objective function.

Again, an alternating minimization method is employed to solve (5.7); i.e., in one step, (5.7) is solved for the image f by fixing the estimated PSF h and the observed image g, and in the other step, (5.7) is solved for the PSF h by fixing the estimated image f and the observed image g. The alternating minimization algorithm solves these two problems iteratively until some stopping criteria are satisfied. The algorithm is summarized in Algorithm 2. We note that there are two main problems in steps 3 and 4. For step 3, we have showed how to solve the image estimation problem; see section 5.2.1. In the following, we will present how to solve the PSF estimation problem in step 4.

Algorithm 2 Bind image deconvolution

1. Input: an initial image f_0, an initial PSF h_0, and an observed image g.
2. Let ϵ_f and ϵ_h be the threshold values for stopping criteria, and $k = 1$.
3. Solve
$$f_k = \text{argmin}_f J(f, h_{k-1}) \equiv \text{argmin}_f \|h_{k-1} \otimes f - g\|_{L^2}^2 + \alpha_1 TV(f).$$
4. Solve
$$h_k = \text{argmin}_h J(f_k, h) \equiv \text{argmin}_h \|h \otimes f_k - g\|_{L^2}^2 + \alpha_2 \|\Delta h\|_{L^2}^2.$$
5. **if** $\frac{\|f_k - f_{k-1}\|_{L^2}}{\|f_{k-1}\|_{L^2}} \leq \epsilon_f$ or $\frac{\|h_k - h_{k-1}\|_{L^2}}{\|h_{k-1}\|_{L^2}} \leq \epsilon_h$, **then**
6. Goto step 10
7. **else**
8. $k = k + 1$
9. Goto step 3
10. **end if**
11. Output: the restored image f_k and the restored PSF h_k.

The restoration of PSF

In step 4 of Algorithm 2, we need to minimize the following problem to restore the PSF h from the given f and g:

$$\min_h \|h \otimes f - g\|_{L^2}^2 + \alpha_2 \|\Delta h\|_{L^2}^2. \tag{5.8}$$

After discretization, we propose to solve (5.8) in frequency domain, i.e.,

$$\min_{\hat{\mathbf{h}}} \|\Gamma \hat{\mathbf{h}} - \hat{\mathbf{g}}\|_2^2 + \alpha_2 \|\Omega \hat{\mathbf{h}}\|_2^2, \tag{5.9}$$

where Γ is a diagonal matrix containing the Fourier transform of f, both vectors $\hat{\mathbf{h}}$ and $\hat{\mathbf{g}}$ are the Fourier transforms of h and g respectively, and Ω is a diagonal matrix which contains the eigenvalues of the discrete Laplacian operator. Since Γ and Ω are diagonal matrices, the problem is easy to solve. The final solution of (5.9) in spatial domain is

$$\mathbf{h} = F^* \hat{\mathbf{h}} = F^* \left[\left(\Gamma^* \Gamma + \alpha_2 \Omega^* \Omega \right)^{-1} \Gamma^* \hat{\mathbf{g}} \right].$$

The regularization parameter α_2 can be determined by the GCV method given as follows.

The selection of regularization parameter

We assume that Ω is invertible and let $\hat{\mathbf{h}}' = \Omega \hat{\mathbf{h}}$. Then the optimization problem (5.9) is equivalent to

$$\min_{\hat{\mathbf{h}}'} \|\Gamma \Omega^{-1} \hat{\mathbf{h}}' - \hat{\mathbf{g}}\|_2^2 + \alpha_2 \|\hat{\mathbf{h}}'\|_2^2.$$

Note that both Γ and Ω are diagonal matrices. With the results in [6, 10], we obtain the GCV function of α_2 for (5.9):

$$GCV_{\mathbf{h}}(\alpha_2) = \frac{n^2 \sum_{i=1}^{n^2} \left(\frac{|[\Omega]_{i,i}|^2 |(\hat{\mathbf{g}})_i|}{|[\Gamma]_{i,i}|^2 + \alpha_2 |[\Omega]_{i,i}|^2} \right)^2}{\left(\sum_{i=1}^{n^2} \frac{|[\Omega]_{i,i}|^2}{|[\Gamma]_{i,i}|^2 + \alpha_2 |[\Omega]_{i,i}|^2} \right)^2}. \tag{5.10}$$

If Ω is not invertible, we can replace Ω by $\Omega + \tau I$ with $\tau > 0$. In this case, the GCV function can be derived similar to (5.10) except $[\Omega]_{i,i}$ is replaced by $[\Omega]_{i,i} + \tau$. We can obtain the limiting GCV function of α_2 by considering that τ tends to zero. The resulting GCV function is just the same as (5.10). The optimal regularization parameter α_2 can be chosen to be the one that minimizes $GCV_{\mathbf{h}}(\alpha_2)$, as (5.6) in section 5.2.1.

5.3 Experimental results

In this section, the proposed non-blind and blind deconvolution algorithms are tested on different test images with different kinds of PSFs and noise levels. We also compare the proposed algorithms with some state-of-the-art algorithms. We use Blurred Signal to Noise Ratio (BSNR) to measure the noise contained in observed images. We then use Signal to Noise Ratio (SNR) and Improvement in Signal to Noise Ratio (ISNR) to measure the quality of restored images. They are defined as follows:

$$\text{SNR} = 20 \log_{10} \left(\frac{\|\mathbf{f} - \text{mean}(\mathbf{f})\|_2}{\|\mathbf{u} - \mathbf{f}\|_2} \right), \quad \text{BSNR} = 20 \log_{10} \left(\frac{\|\mathbf{g}\|_2}{\|\mathbf{w}\|_2} \right),$$

and

$$\text{ISNR} = 20 \log_{10} \left(\frac{\|\mathbf{f} - \mathbf{g}\|_2}{\|\mathbf{f} - \mathbf{u}\|_2} \right),$$

where $\mathbf{f}$, $\mathbf{g}$, $\mathbf{u}$, and $\mathbf{w}$ are the vectors of the original image, the observed image, the recovered image, and the noise, respectively. The notation mean($\mathbf{f}$) denotes the mean of $\mathbf{f}$.

5.3.1 Experiments on the non-blind deconvolution method

The "Cameraman" image is blurred by a Gaussian PSF of size 7 with the standard derivation of 5. The "Barbara" image is blurred by out-of-focus PSFs with radii of 3 and 7 respectively. The "CarNo" image is blurred by motion PSFs of length 9 and 15 pixels respectively. Then the Gaussian noise is added to each blurred image. We list the test results and compare them with the fast-TV method [9] in Table 5.1. We can see from the table that our restoration results are close to those by the fast-TV method. Note that there are two parameters in the fast-TV method. We list the best results by finding the "optimal" parameters, i.e., we test 10000 combinations of these two parameters and then report the best ones in Table 5.1. We remark that it is quite time-consuming to find the "optimal" parameters, and the original image is assumed to be known to help selecting the best result for each test. We display some restored images for comparison in Figure 5.1. It can be seen that the restored images by Algorithm 1 are quite well visually.

Table 5.1: Comparison on test results (SNR).

Image	PSF/BSNR	Algorithm 1	Fast-TV[9]
Cameraman	Gaussian(7,5)/40	17.47	18.13
Cameraman	Gaussian(14,5)/30	11.65	12.10
Barbara	Out-of-focus(3)/40	17.40	17.99
Barbara	Out-of-focus(7)/30	13.30	13.49
CarNo	Motion(9)/40	20.41	22.03
CarNo	Motion(15)/30	15.65	16.75

(a) Cameraman, Gaussian (7,5) PSF, BSNR=40dB.

(b) Cameraman, Gaussian (7,5) PSF, BSNR=40dB.

(c) Barbara, out-of-focus (3) PSF, BSNR=40dB.

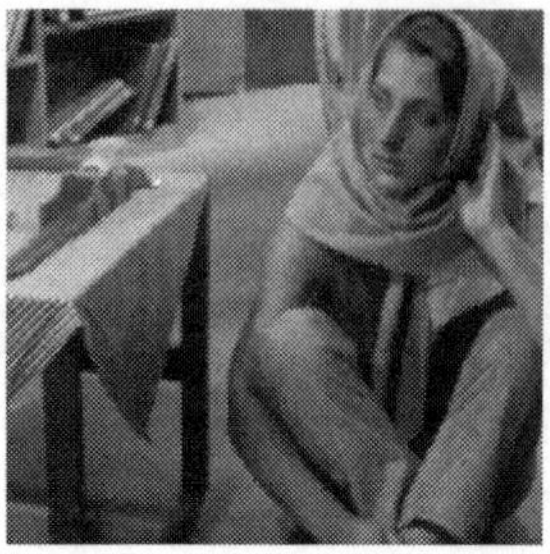

(d) Barbara, out-of-focus (3) PSF, BSNR=40dB.

(e) CarNo, motion(9) PSF, BSNR=40dB.

(f) CarNo, motion(9) PSF, BSNR=40dB.

Figure 5.1: Comparison on some restored images: left by the proposed method; right by the fast-TV[9].

Table 5.2: Comparison on Gaussian PSF with variance 9 (ISNR).

BSNR	Method	Lena	Cameraman	Shepp-Logan
40dB	Algorithm 2	3.54	3.57	3.47
40dB	TV1	2.53	1.82	3.07
40dB	TV2	2.95	1.73	3.36
20dB	Algorithm 2	2.60	3.56	3.75
20dB	TV1	2.62	1.70	2.47
20dB	TV2	-32.50	-40.89	-23.88

Table 5.3: Comparison on Gaussian PSF with variance 5 (ISNR).

BSNR	Method	Lena	Cameraman	Shepp-Logan
40dB	Algorithm 2	3.75	4.97	5.07
40dB	TV1	3.19	1.66	2.05
40dB	TV2	3.29	2.49	3.79
20dB	Algorithm 2	3.32	4.63	4.84
20dB	TV1	1.39	1.43	2.09
20dB	TV2	-45.20	-42.54	-26.00

5.3.2 Experiments on the blind deconvolution method

In this section, we compare Algorithm 2 with TV1 and TV2 in [2]. In their methods, they used the TV function as the image prior, and a simultaneous autoregressive model as the PSF prior in the framework of Bayesian formulation. Then a variational method is used to approximate the posterior distribution by a product of distributions using the Kullback-Leibler divergence for estimating model parameters, image, and PSF simultaneously. The test images are "Cameraman", "Lena", and "Shepp-Logan". The original images are blurred with a Gaussian PSF having variances 9 and 5 respectively. Then Gaussian noises BSNR=40dB and 20dB are added to the blurred images respectively. In Tables 5.2 and 5.3, we list the restoration results of the TV1, the TV2, and the proposed method. We see from tables that the proposed method is better than TV1 and TV2, especially on the "Cameraman" image.

Next we compare the proposed method with the recent variational Bayesian blind deconvolution algorithm (StStSt) [16] and the TV blind deconvolution algorithm [4]. Here we blur the "Cameraman" image with the Gaussian and the square PSFs respectively. We then add noises at different levels to blurred images, the same test setting as used in [16] by StStSt and the TV blind deconvolution method. The details and test results are shown in Table 5.4. We can see that Algorithm 2 performs better.

Table 5.4: Comparison on "Cameraman" image with different PSFs and noise levels (SNR).

BSNR	PSF	Algorithm 2	StStSt in [16]	Method in [4]
40dB	Guassian ($\sigma^2 = 5$)	4.97	2.82	1.32
40dB	Square (7×7)	8.40	8.31	4.06
20dB	Guassian ($\sigma^2 = 5$)	4.63	1.57	1.17
20dB	Square (7×7)	2.97	2.69	2.56

5.4 Concluding remarks

In this chapter, we study and develop two TV based algorithms for non-blind and blind deconvolution problems respectively. The main contribution is that the GCV method is studied on the regularization parameter problems within the alternating minimization scheme. Experimental results demonstrate the proposed algorithms are competitive with other existing state-of-the-art methods in terms of quantitative measures and visual effect.

References for Chapter 5

[1] S. D. Babacan, R. Molina, and A. K. Katsaggelos, *Parameter estimation in TV image restoration using variation distribution approximation*, IEEE Trans. Image Process., 17 (2008), pp. 326–339.

[2] S. D. Babacan, R. Molina, and A. K. Katsaggelos, *Variational Bayesian blind deconvolution using a total variation prior*, IEEE Trans. Image Process., 18 (2009) pp. 12–26.

[3] A. Chambolle, *An algorithm for total variation minimization and applications*, J. Math. Imaging Vision, 20 (2004), pp. 89–97.

[4] T. F. Chan and C. K. Wong, *Total variation blind deconvolution*, IEEE Trans. Image Process., 7 (1998), pp. 370–375.

[5] R. Courant, *Variational methods for the solution of problems with equilibrium and vibration*, Bull. Amer. Math. Soc., 49 (1943), pp. 1–23.

[6] G. Golub, M. Heath, and G. Wahba, *Generalized cross-validation as a method for choosing a good ridge parameter*, Technometrics, 21 (1979), pp. 215–223.

[7] G. Golub and U. Matt, *Generalized cross-validation for large scale problems*, J. Comput. Graph. Stat., 6 (1997), pp. 1–34.

[8] X. Guo, F. Li, and M. Ng, *A fast l_1-TV algorithm for image restoration*, SIAM J. Sci. Comput., 31 (2009), pp. 2322–2341.

[9] Y. Huang, M. Ng, and Y. Wen, *A fast total variation minimization method for image restoration*, SIAM J. Multiscale Model. Simul., 7 (2008), pp. 774–795.

[10] M. Kilmer and D. O'leary, *Choosing regularization parameters in iterative methods for ill-posed problems*, SIAM. J. Matrix Anal. Appl., 22 (2001), pp. 1204–1221.

[11] Y. Lin and B. Wohlberg, *Application of the UPRE method to optimal parameter selection for large scale regularization problems,* IEEE Southwest Symposium on Image Analysis and Interpretation, Santa Fe, NM, USA, (2008), pp. 89–92.

[12] R. Molina, J. Mateos, and A. K. Katsaggelos, *Blind deconvolution using a variational approach to parameter, image and blur estimation*, IEEE Trans. Image Process., 15 (2006), pp. 3715–3727.

[13] M. Ng, *Iterative Methods for Toeplitz Systems*, Oxford University Press, Oxford, 2004.

[14] S. Reeves and R. Mersereau, *Blur identification by the method of generalized crossvalidation*, IEEE Trans. Image Process., 1 (1992), pp. 301–311.

[15] L. Rudin, S. Osher, and E. Fatemi, *Nonlinear total variation based noise removal algorithms*, Physica D: Nonlinear Phenomena, 60 (1992), pp. 259–268.

[16] D. Tzikas, A. Likas, and N. Galatsanos, *Variational Bayesian sparse kernel-based blind image deconvolution with student's-t priors*, IEEE Trans. Image Process., 18 (2009), pp. 753–764.

[17] Y. Wang, J. Yang, W. Yin, and Y. Zhang, *A new alternating minimization algorithm for total variation image reconstruction*, SIAM J. Imaging Sciences, 1 (2008), pp. 248–272.

Chapter 6

Matrices and the LLL Algorithm

Franklin T. Luk [1] **San-zheng Qiao** [2]

Abstract

We present a matrix interpretation of the LLL algorithm for lattice basis reduction.

6.1 Introduction

The LLL algorithm [3] is a lattice basis reduction method. It was originally presented in the context of factorizing polynomials with rational coefficients. Since the publication of their paper in 1982, a wide range of applications have been found [5]. In this chapter, we present a matrix interpretation of the algorithm, including the proof of the justification of the LLL-reduced basis. We hope that our interpretation can benefit readers in various application fields. The chapter is organized as follows. After an introduction to lattices and bases in section 6.2, we present the definition of LLL-reduced basis in section 6.3. Then in section 6.4, we describe the LLL algorithm in a matrix form.

6.2 Lattices and bases

Given a set of linearly independent vectors $\mathbf{a}_i \in \mathbb{R}^m$, $i = 1, 2, \ldots, n$, $m \geq n$, the integer combinations of the vectors are represented by discrete points in $\mathbb{R}^m$. Those discrete points form a *lattice* and the vectors $\mathbf{a}_i$ form a *basis* for the lattice. Let $A = [\mathbf{a}_1 \ \mathbf{a}_2 \ \ldots \ \mathbf{a}_n] \in \mathbb{R}^{m\times n}$ in term of matrix, then A is of full column rank and the lattice can be expressed as

$$L = \{A\mathbf{z}, \quad \mathbf{z} \in \mathbb{Z}^n\},$$

[1]Department of Mathematics, Hong Kong Baptist University, Kowloon Tong, Hong Kong.

[2]Department of Computing and Software, McMaster University, Hamilton, Ontario, Canada L8S 4K1 (email: qiao@mcmaster.ca).

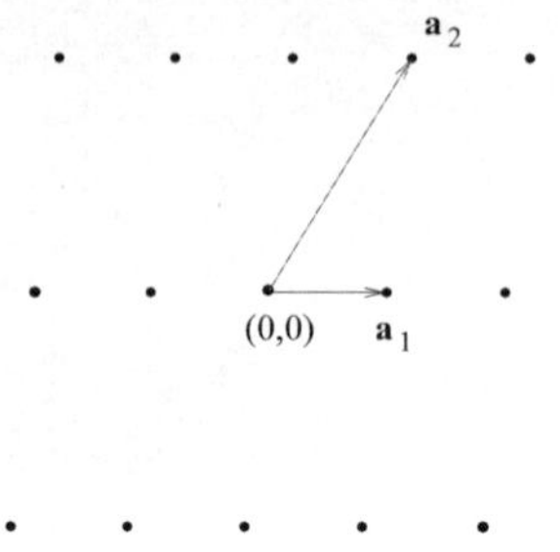

Figure 6.1: The column vectors $\mathbf{a}_1$ and $\mathbf{a}_2$ of A in (6.1) and the lattice points generated by them.

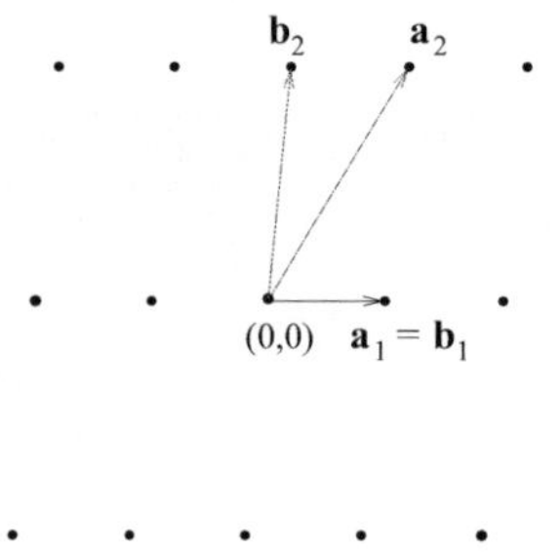

Figure 6.2: The columns of A in (6.1) and those of B in (6.2) and the lattice.

where $\mathbb{Z}^n$ denotes the set of all integer vectors. Figure 6.1 depicts the lattice generated by the columns $\mathbf{a}_1$ and $\mathbf{a}_2$ of A:

$$A = [\mathbf{a}_1 \ \ \mathbf{a}_2] = \begin{bmatrix} 1.0 & 1.2 \\ 0 & 2.0 \end{bmatrix}. \tag{6.1}$$

If $Z \in \mathbb{Z}^{n \times n}$ is an integer matrix, then the columns of $B = AZ$ generate a sublattice of L. Thus when Z is nonsingular and its inverse is also an integer matrix, then both A and B generate the same lattice L. For example,

$$B = [\mathbf{b}_1 \ \ \mathbf{b}_2] = \begin{bmatrix} 1.0 & 0.2 \\ 0 & 2.0 \end{bmatrix} = AZ, \quad Z = \begin{bmatrix} 1 & -1 \\ 0 & 1 \end{bmatrix}, \tag{6.2}$$

generates the same lattice as shown in Figure 6.2.

In general, two matrices A and B generate the same lattice if and only if $B = AZ$ for some nonsingular integer matrix Z whose inverse is also an integer matrix. It follows from the following definition that a nonsingular integer matrix has an integer inverse if and only if it is unimodular.

Definition 6.1 (Unimodular) *A nonsingular integer matrix M is called unimodular if* $\det(M) = \pm 1$.

Obviously, from the definition, $|\det(A)| = |\det(AM)|$, for any unimodular M. Consequently, the quantity

$$\det(L) = |\det(A)|$$

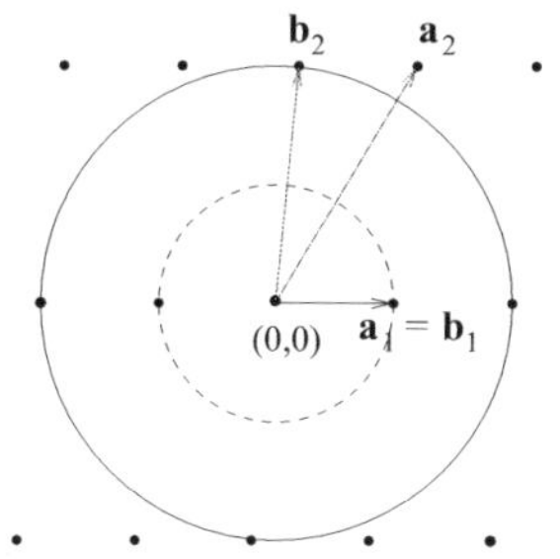

Figure 6.3: A geometric interpretation of reduced $B = [\mathbf{b}_1 \;\; \mathbf{b}_2]$ in (6.2) and not reduced $A = [\mathbf{a}_1 \;\; \mathbf{a}_2]$ in (6.1).

is independent of the choice of basis for L and thus defined as the *determinant* of a lattice L.

If the columns of A form a basis for a lattice L, for any unimodular matrix Z, the columns of AZ also form a basis for L. That means a lattice can have multiple bases. Are there some bases more desirable than others? Figure 6.2 depicts that the basis $\{\mathbf{b}_1, \mathbf{b}_2\}$ is "better" than $\{\mathbf{a}_1, \mathbf{a}_2\}$, since the basis vectors $\mathbf{b}_1$ and $\mathbf{b}_2$ are short and "nearly" orthogonal. This introduces the concept of reduced basis, illustrated as follows.

Let L be the lattice generated by the columns of A in (6.1) or B in (6.2), then for any integer vector $\mathbf{z}$, $A\mathbf{z}$ or $B\mathbf{z}$ is a lattice vector in L. Since short basis vectors are desirable, we first find a shortest nonzero lattice vector. Figure 6.3 shows that $\mathbf{b}_1 = B\mathbf{e}_1$, where $\mathbf{e}_1 = (1, 0)^T$, is a shorted nonzero lattice vector. This figure also shows that $\mathbf{b}_2 = B\mathbf{e}_2$, where $\mathbf{e}_2 = (0, 1)^T$, is a shortest lattice vector that is linearly independent of $\mathbf{b}_1$. In contrast, $\mathbf{a}_2$ is not a shortest lattice vector that is linearly independent of $\mathbf{a}_1 = \mathbf{b}_1$. In this example, $\mathbf{b}_1$ and $\mathbf{b}_2$ are successively shortest basis vectors for L. In general, we have the following definitions of Minkowski minima and Minkowski-reduced basis.

Definition 6.2 (Minkowski minima) [1, p.201] *Using the Euclidean norm as a distance function, we say that λ_k, $1 \leq k \leq n$, is the kth successive minimum with respect to a lattice L, if λ_k is the lower bound of the radius λ of the sphere*

$$\|B\mathbf{z}\|_2 \leq \lambda$$

that contains k linearly independent lattice points.

In other words, λ_k is the lower bound for $\max(\|\mathbf{x}_1\|_2, \|\mathbf{x}_2\|_2, \ldots, \|\mathbf{x}_k\|_2)$ over all sets of linearly independent lattice points $\mathbf{x}_1, \mathbf{x}_2, \ldots, \mathbf{x}_k$. Note that originally, the Minkowski minima are defined in terms of $\|\mathbf{x}_i\|_2^2$ in the context of quadratic form. Here, for simplicity, we define the Minkowski minima in terms of $\|\mathbf{x}_i\|_2$.

Definition 6.3 (Minkowski-reduced) *A lattice basis $\{\mathbf{b}_1, \mathbf{b}_2, \ldots, \mathbf{b}_n\}$ is called Minkowski-reduced if $\|\mathbf{b}_i\|_2 = \lambda_i$ for $i = 1, 2, \ldots, n$.*

A lattice may not have a Minkowski-reduced basis, first noticed by Korkine and Zolotarev in the nineteenth century [5, p.33]. For example, consider the lattice basis formed by the columns of the following matrix:

$$\begin{bmatrix} 2 & 0 & 0 & 0 & 1 \\ 0 & 2 & 0 & 0 & 1 \\ 0 & 0 & 2 & 0 & 1 \\ 0 & 0 & 0 & 2 & 1 \\ 0 & 0 & 0 & 0 & 1 \end{bmatrix}.$$

All its Minkowski minima equal two, since $(0,0,0,0,2)^T$ is a lattice vector linearly independent of the first four columns and its length is two. However, this vector and the first four columns of the above matrix do not form a basis for the lattice. Actually, they form a basis for a proper sublattice.

6.3 LLL-reduced bases

Suppose that the columns $\mathbf{b}_1, \mathbf{b}_2, \ldots, \mathbf{b}_n$ of matrix B are linearly independent and the application of the Gram-Schmidt orthogonalization yields the decomposition

$$B = Q^* U, \tag{6.3}$$

where Q^* has orthogonal columns $\mathbf{q}_i^*$ and $U = [u_{i,j}]$ is upper triangular with unit diagonal, i.e., $u_{i,j} = 0$ when $i > j$ and $u_{i,i} = 1$. Setting the diagonal matrix $D = \mathrm{diag}(d_1, d_2, \ldots, d_n)$, where $d_i = \|\mathbf{q}_i^*\|_2$, we further decompose Q^*:

$$Q^* = QD,$$

then $Q = [\mathbf{q}_1 \ \ldots \ \mathbf{q}_n]$ has orthonormal columns. Let $R = [r_{i,j}] = DU$, then we have the QR decomposition [2]:

$$B = QR. \tag{6.4}$$

Thus we have the following relations:

$$\mathbf{q}_i = \mathbf{q}_i^* / \|\mathbf{q}_i^*\|_2, \qquad r_{i,i} = d_i = \|\mathbf{q}_i^*\|_2, \qquad r_{i,j} = r_{i,i} u_{i,j} = d_i u_{i,j}. \tag{6.5}$$

The QR decomposition can be obtained in ways other than the Gram-Schmidt method, such as Householder transformations or Givens transformations [2].

In [3], an LLL-reduced basis is defined as follows. The columns $\mathbf{b}_i$ of a full column rank matrix B form an *LLL-reduced* basis for a lattice if the matrices Q^* and U in the decomposition (6.3) satisfy

$$|u_{i,j}| \le 1/2, \quad j > i,$$

and

$$\|\mathbf{q}_i^*\|_2^2 + u_{i-1,i}^2 \|\mathbf{q}_{i-1}^*\|_2^2 \ge \omega \|\mathbf{q}_{i-1}^*\|_2^2,$$

where $\omega \in (0.25, 1.0)$ is a parameter associated with the definition. If a basis is LLL-reduced with respect to ω_1, then it is LLL-reduced with respect to any $\omega \le \omega_1$. In other words, the larger the value of ω, the more stringent the definition.

In terms of the QR decomposition (6.4), using the relations in (6.5), we have the following definition [4].

Definition 6.4 (LLL-reduced) *The columns* $\mathbf{b}_i$ *of a full column rank matrix* B *form an* LLL-reduced *basis for a lattice if the entries of the upper triangular matrix* R *in the decomposition* (6.4) *satisfy*

$$|r_{i,j}| \le |r_{i,i}|/2, \quad j > i, \tag{6.6}$$

and

$$r_{i,i}^2 + r_{i-1,i}^2 \ge \omega\, r_{i-1,i-1}^2 \tag{6.7}$$

for a given $\omega \in (0.25, 1.0)$.

Since the conditions in the above definition solely depend on R, we may simply call R LLL-reduced. In the rest of the chapter, we adopt this QR decomposition approach.

To justify the above definition of reduced bases, it is shown in [3] that if $\mathbf{b}_1, \mathbf{b}_2, \ldots, \mathbf{b}_n$ form an LLL-reduced basis for a lattice, then

$$\eta^{1-i}\lambda_i^2 \le \|\mathbf{b}_i\|_2^2 \le \eta^{n-1}\lambda_i^2, \tag{6.8}$$

where $\eta = (\omega - 1/4)^{-1}$ and $\lambda_1, \lambda_2, \ldots, \lambda_n$ are the Minkowski successive minima. This says that the lengths of an LLL-reduced basis vectors are bounded by the corresponding Minkowski minima. However, note that η can be arbitrarily large as ω approaches $1/4$. In particular, when $\omega = 3/4$, then $\eta = 2$ and

$$2^{1-i}\lambda_i^2 \le \|\mathbf{b}_i\|_2^2 \le 2^{n-1}\lambda_i^2.$$

To prove the first inequality in (6.8), we first show that

$$\|\mathbf{b}_j\|_2^2 \le \eta^{i-1} r_{i,i}^2, \quad \text{for } 1 \le j \le i \le n. \tag{6.9}$$

Indeed, from the conditions (6.6) and (6.7), and the definition of η, we have

$$r_{i,i}^2 \ge (\omega - r_{i,i-1}^2/r_{i-1,i-1}^2) r_{i-1,i-1}^2 \ge \eta^{-1} r_{i-1,i-1}^2,$$

implying that

$$r_{j,j}^2 \le \eta^{i-j} r_{i,i}^2, \quad \text{for } 1 \le j \le i \le n. \tag{6.10}$$

On the other hand, from the decomposition (6.4), the condition (6.6), the inequality (6.10), and the definition of η, which implies that $\eta > 4/3$, we get

$$\begin{aligned}
\|\mathbf{b}_j\|_2^2 &= r_{1,j}^2 + r_{2,j}^2 + \cdots + r_{j-1,j}^2 + r_{j,j}^2 \le \frac{1}{4}(r_{1,1}^2 + r_{2,2}^2 + \cdots + r_{j-1,j-1}^2) + r_{j,j}^2 \\
&\le \frac{1}{4}(\eta^{j-1} + \eta^{j-2} + \cdots + \eta) r_{j,j}^2 + r_{j,j}^2 = \left[\frac{\eta}{4(\eta-1)}(\eta^{j-1} - 1) + 1\right] r_{j,j}^2 \\
&< \eta^{j-1} r_{j,j}^2 \le \eta^{i-1} r_{i,i}^2,
\end{aligned} \tag{6.11}$$

which proves (6.9). It then follows that

$$\|\mathbf{b}_i\|_2^2 \ge r_{i,i}^2 \ge \eta^{1-i}\|\mathbf{b}_j\|_2^2,$$

for $j = 1, \ldots, i$. Since λ_i is the ith successive minimum, the first inequality in (6.8) is proved.

The upper bound in (6.9) is almost attainable. Consider the matrix

$$B = \begin{bmatrix} 1 & & & & \\ & \eta^{-1/2} & & 0 & \\ & & \eta^{-2/2} & & \\ & 0 & & \ddots & \\ & & & & \eta^{-(n-1)/2} \end{bmatrix} \begin{bmatrix} 1 & \frac{1}{2} & \frac{1}{2} & \cdots & \frac{1}{2} \\ & 1 & \frac{1}{2} & \cdots & \frac{1}{2} \\ & & 1 & \cdots & \frac{1}{2} \\ & 0 & & \ddots & \vdots \\ & & & & 1 \end{bmatrix}. \tag{6.12}$$

In this case, following the derivation of (6.11), we have

$$\|\mathbf{b}_j\|_2^2 = \left[\frac{\eta}{4(\eta-1)}(\eta^{j-1}-1)+1\right] r_{j,j}^2,$$

which equals $\frac{1}{2}(\eta^{j-1}+1)r_{j,j}^2$ when $\eta = 2$, i.e., $\omega = 3/4$. It is almost $\eta^{j-1}r_{j,j}^2$ and exactly $\eta^{j-1}r_{j,j}^2$ when $j = 1$. Also in the case of (6.12), $r_{j,j}^2 = \eta^{i-j}r_{i,i}^2$, for $1 \le j \le i \le n$. Thus, the upper bound for $\|\mathbf{b}_j\|_2^2$ in (6.9) is almost attainable.

Note that the determinant $\det(L) = \prod_{i=1}^{n} |r_{i,i}|$, it follows from (6.9) that

$$\prod_{i=1}^{n} \|\mathbf{b}_i\|_2 \le \eta^{\frac{n(n-1)}{4}} \det(L),$$

and, from the above argument, this upper bound is almost attainable. On the other hand,

$$\det(L) \le \prod_{i=1}^{n} \|\mathbf{b}_i\|_2,$$

since $\|\mathbf{b}_i\|_2 \ge |r_{i,i}|$.

Setting $j = 1$ in (6.9), we have $\|\mathbf{b}_1\|_2 \le \eta^{\frac{i-1}{2}} |r_{i,i}|$, for $1 \le i \le n$, and then

$$\|\mathbf{b}_1\|_2^n \le \eta^{\frac{n(n-1)}{4}} \det(L),$$

which implies that

$$\|\mathbf{b}_1\|_2 \le \eta^{\frac{n-1}{4}} \det(L)^{\frac{1}{n}}.$$

This upper bound is attainable in the case of (6.12), since $\|\mathbf{b}_1\|_2^2 = r_{1,1}^2 = \eta^{i-1}r_{i,i}^2$, for $1 \le i \le n$.

Recall that the ith successive minimum λ_i is the lower bound for

$$\max(\|\mathbf{x}_1\|_2, \|\mathbf{x}_2\|_2, \ldots, \|\mathbf{x}_i\|_2),$$

for any set of linearly independent lattice points $\mathbf{x}_1, \mathbf{x}_2, \ldots, \mathbf{x}_i$. To prove the second inequality in (6.8), it suffices to show that for any set of linearly independent lattice points $\mathbf{x}_1, \mathbf{x}_2, \ldots, \mathbf{x}_i$,

$$\|\mathbf{b}_j\|_2^2 \le \eta^{n-1} \max(\|\mathbf{x}_1\|_2^2, \|\mathbf{x}_2\|_2^2, \ldots, \|\mathbf{x}_i\|_2^2), \tag{6.13}$$

for $j = 1, 2, \ldots, i$.

Let $\mathbf{x}_j = B\mathbf{z}_j$ be a lattice point, where $\mathbf{z}_j = [z_{k,j}] \in \mathbb{Z}^n$. Suppose that $z_{k_j,j}$ is the leading nonzero entry of $\mathbf{z}_j$, i.e., $|z_{k_j,j}| \ge 1$ and $z_{k_j+1,j} = \cdots = z_{n,j} = 0$. We may order $\mathbf{x}_1, \mathbf{x}_2, \ldots, \mathbf{x}_i$ so that $k_1 \le k_2 \le \cdots \le k_i$. Then we have $k_j \ge j$ for all $1 \le j \le i$, because if $k_j < j$ for some j, then $1 \le k_1 \le k_2 \le \cdots \le k_j < j$, implying that $\mathbf{x}_1, \mathbf{x}_2, \ldots, \mathbf{x}_j$ could be generated by $\mathbf{b}_1, \mathbf{b}_2, \ldots, \mathbf{b}_{j-1}$, contradicting the linear independence of $\mathbf{x}_i$. Thus from (6.9), we have

$$\|\mathbf{b}_j\|_2^2 \le \eta^{k_j-1} r_{k_j,k_j}^2 \le \eta^{n-1} r_{k_j,k_j}^2.$$

Finally, since $|z_{k_j,j}| \ge 1$, it follows from the above inequality that

$$\|\mathbf{b}_j\|_2^2 \le \eta^{n-1} z_{k_j,j}^2 r_{k_j,k_j}^2 \le \eta^{n-1} \|\mathbf{x}_j\|_2^2,$$

for all $1 \le j \le i$. This proves (6.13).

6.4 The LLL algorithm

Given a matrix A, whose columns form a basis for a lattice L, the LLL algorithm computes a matrix B, whose columns form a reduced basis for L in the sense of Definition 6.4.

As a preprocessing, the algorithm first triangularizes A through orthogonal transformations, for example, the QR decomposition (6.4):

$$A = Q_1 S,$$

where Q_1 has orthonormal columns and $S = [s_{i,j}]$ is upper triangular. Alternatively [3], the Gram-Schmidt orthogonalization (6.3) can be applied to A to obtain the decomposition

$$A = Q_1 D U, \tag{6.14}$$

where $D = \mathrm{diag}(d_1, d_2, \ldots, d_n)$ and $U = [u_{i,j}]$ is upper triangular with unit diagonal. In particular, U is a row scaled S, i.e., $S = DU$. For efficiency, instead of d_i, the values of d_i^2 are computed to avoid square root operations [3]. Thus the triangularization applies orthogonal transformations to the lattice basis vectors.

After the initial triangularization, the LLL algorithm computes the QRZ decomposition

$$S = Q_2 R Z^{-1},$$

where Q_2 is orthogonal, Z is unimodular, and R is reduced as defined in Definition 6.4. The parameter ω in (6.7) controls the termination of the algorithm. The smaller the ω is, the sooner the algorithm terminates. Let $Q = Q_1 Q_2$, then $A = QRZ^{-1}$. Thus the columns of $B = QR = AZ$ form an LLL-reduced basis for L. Alternatively, if the Gram-Schmidt orthogonalization is used in the initial stage, then R is given in the form $R = DU$. For efficiency, the LLL algorithm updates d_i^2 to avoid square-root operation.

Now, we describe the second stage of the algorithm. Suppose that R is initialized as S and modified in the algorithm. When the algorithm terminates, R is LLL-reduced. The second stage essentially consists of two procedures that impose the two conditions (6.6) and (6.7). When $|r_{i,j}| > |r_{i,i}|/2$, for some i and $j > i$, the following procedure is invoked to enforce the condition (6.6).

Procedure 6.1 (`Decrease`(i, j)) *Given R and Z, calculate $\gamma = \lceil r_{i,j}/r_{i,i} \rfloor$ ($\lceil a \rfloor$ denotes an integer that is closest to a), form $Z_{ij} = I_n - \gamma \mathbf{e}_i \mathbf{e}_j^{\mathrm{T}}$* ($\mathbf{e}_i$ and I_n denote the ith unit vector and the identity matrix of order n respectively), *and apply Z_{ij} to both R and Z:*

$$R \leftarrow R Z_{ij} \quad \text{and} \quad Z \leftarrow Z Z_{ij}.$$

Thus, if $|r_{i,i}| < 2|r_{i,j}|$ in the current R, then in the updated R, we have $|r_{i,i}| \geq 2|r_{i,j}|$ satisfying the condition (6.6).

In terms of D and U in (6.14), in [3], if $|u_{i,j}| > 1/2$, we calculate $\gamma = \lceil u_{i,j} \rfloor$, form $Z_{ij} = I_n - \gamma \mathbf{e}_i \mathbf{e}_j^{\mathrm{T}}$, and apply Z_{ij} to both U and Z:

$$U \leftarrow U Z_{ij} \quad \text{and} \quad Z \leftarrow Z Z_{ij}.$$

The condition (6.7) is enforced by the following procedure [4]:

Procedure 6.2 (`SwapRestore(`i`)`) *Given R, Z, and Q, compute the plane reflection*

$$J_i = \begin{bmatrix} c & s \\ s & -c \end{bmatrix}$$

such that

$$J_i \begin{bmatrix} r_{i-1,i-1} & r_{i-1,i} \\ 0 & r_{i,i} \end{bmatrix} P = \begin{bmatrix} \hat{r}_{i-1,i-1} & \hat{r}_{i-1,i} \\ 0 & \hat{r}_{i,i} \end{bmatrix}, \quad \textit{where} \quad P = \begin{bmatrix} 0 & 1 \\ 1 & 0 \end{bmatrix},$$

and apply

$$Q_i = \begin{bmatrix} I_{i-2} & \mathbf{0} & \mathbf{0} \\ \mathbf{0} & J_i & \mathbf{0} \\ \mathbf{0} & \mathbf{0} & I_{n-i} \end{bmatrix} \quad \textit{and permutation} \quad \Pi_i = \begin{bmatrix} I_{i-2} & \mathbf{0} & \mathbf{0} \\ \mathbf{0} & P & \mathbf{0} \\ \mathbf{0} & \mathbf{0} & I_{n-i} \end{bmatrix} \quad (2 \leq i \leq n)$$

to R, Z, and Q:

$$R \leftarrow Q_i R \Pi_i, \qquad Z \leftarrow Z \Pi_i, \qquad Q \leftarrow Q Q_i.$$

The above procedure swaps the columns $i-1$ and i of R and then restores its upper triangular structure. If in the current R, we have $r_{i,i}^2 + r_{i-1,i}^2 < \omega\, r_{i-1,i-1}^2$, where

$$|r_{i-1,i}| \leq |r_{i-1,i-1}|/2,$$

then in the updated R, we have $r_{i,i}^2 + r_{i-1,i}^2 \geq \omega\, r_{i-1,i-1}^2$.

In terms of D and U in (6.14), in [3], the nonsingular matrix

$$X_i = \begin{bmatrix} u_{i-1,i} & 1 - u_{i-1,i}\xi \\ 1 & -\xi \end{bmatrix}, \quad \text{where } \xi = \frac{u_{i-1,i} d_{i-1}^2}{d_i^2 + u_{i-1,i}^2 d_{i-1}^2},$$

is constructed so that

$$X_i^{-1} \begin{bmatrix} u_{i-1,i} \\ 1 \end{bmatrix} = \begin{bmatrix} 1 \\ 0 \end{bmatrix}.$$

Similar to the reflection J_i in Procedure 6.2, X_i^{-1} restores the structure of U after a column permutation. The corresponding diagonal elements in D are updated by

$$\begin{bmatrix} d_{i-1}^2 \\ d_i^2 \end{bmatrix} \leftarrow \begin{bmatrix} u_{i-1,i}^2 & 1 \\ (1 - u_{i-1,i}\xi)^2 & \xi^2 \end{bmatrix} \begin{bmatrix} d_{i-1}^2 \\ d_i^2 \end{bmatrix}.$$

Thus the algorithm in [3] is square-root free.

The following is the second stage of the LLL basis reduction algorithm using the QR decomposition approach. See [4] for details.

Algorithm 6.1 (Matrix LLL Algorithm) *Given an $n \times n$ nonsingular upper triangular $R = [r_{i,j}]$ and a parameter ω, where $0.25 < \omega < 1$, this algorithm computes an orthogonal matrix Q and an integer unimodular matrix Z. The algorithm overwrites R, so that the new upper triangular matrix R equals $Q^{\mathrm{T}}RZ$ and is LLL-reduced.*

```
set Z ← I and Q ← I;
k ← 2;
while k ≤ n
    if |r_{k-1,k}/r_{k-1,k-1}| > 1/2
            Decrease(k − 1, k);
    endif
    if r_{k,k}^2 + r_{k-1,k}^2 < ω r_{k-1,k-1}^2
            SwapRestore(k);
            k ← max(k − 1, 2);
    else
        for i = k − 2 downto 1
            if |r_{i,k}/r_{i,i}| > 1/2
                Decrease(i, k);
            endif
        endfor
        k ← k + 1;
    endif
endwhile
```

References for Chapter 6

[1] J. W. S. Cassels, *An Introduction to the Geometry of Numbers*, Second Printing, Springer-Verlag, Berlin, Heidelberg, 1997.

[2] G. H. Golub and C. F. Van Loan, *Matrix Computations*, Third Edition, The Johns Hopkins University Press, Baltimore, 1996.

[3] A. K. Lenstra, H. W. Lenstra, Jr., and L. Lovász, *Factorizing polynomials with rational coefficients*, Mathematicsche Annalen, 261 (1982), pp. 515–534.

[4] F. T. Luk and D. M. Tracy, *An improved LLL algorithm*, Linear Algebra and its Applications, 428 (2008), pp. 441–452.

[5] P. Q. Nguyen and B. Vallée (editors), *The LLL Algorithm: Survey and Applications*, Information Security and Cryptography, Springer, Heidelberg, Dordrecht, London, New York, 2010.

Chapter 7

A Fast Nonconvex Nonsmooth Minimization Method for Image Restoration and Reconstruction

Mila Nikolova [1]   Michael K. Ng [2]   Chi-pan Tam [3]

Abstract

Nonconvex nonsmooth regularization has many advantages over convex regularization for restoring images with neat edges. However, its practical interest used to be limited by the difficulty of the computational stage which requires a nonconvex nonsmooth minimization. In this chapter, we study a fast nonconvex nonsmooth minimization method for image restoration and reconstruction. The main aim of this chapter is to develop a fast minimization algorithm to solve the nonconvex nonsmooth minimization problem. Our experimental results show the effectiveness and efficiency of the proposed method.

7.1 Introduction

Digital image restoration and reconstruction play important roles in various applied areas, such as medical and astronomical imaging, film restoration, image and video coding, and so on [3, 8, 23, 25, 36]. We focus on the most common data production model, where the observed data $\mathbf{g} \in \mathbb{R}^q$ is related to the underlying $n \times m$ image which is rearranged into a vector $\mathbf{f} \in \mathbb{R}^p$ with $p = mn$, and according to

$$\mathbf{g} = H\mathbf{f} + \mathbf{n}, \tag{7.1}$$

[1]Centre de Mathématiques et de Leurs Applications, ENS de Cachan, 61 av. President Wilson, 94235 Cachan Cedex, France (email: nikolova@cmla.ens-cachan.fr).

[2]Centre for Mathematical Imaging and Vision and Department of Mathematics, Hong Kong Baptist University, Kowloon Tong, Hong Kong (email: mng@math.hkbu.edu.hk). The research of this author is supported in part by Hong Kong Research Grants Council Grants and HKBU FRGs.

[3]Centre for Mathematical Imaging and Vision and Department of Mathematics, Hong Kong Baptist University, Kowloon Tong, Hong Kong (email: cptam@math.hkbu.edu.hk).

where $\mathbf{n}$ accounts for the perturbations and H is a $q \times p$ matrix representing, for instance, optical blurring, distortion wavelets in seismic imaging and nondestructive evaluation, a Radon transform in X-ray tomography, and a Fourier transform in diffraction tomography. Usually the information provided by the forward model (7.1) alone is not sufficient to find an acceptable solution; say, $\hat{\mathbf{f}}$. Prior information on the underlying image is needed to restore a convenient $\hat{\mathbf{f}}$, which is close to the data production model (7.1) and satisfies some prior requirements. A flexible means to define such a solution is regularization [1, 4, 11, 14]. The solution $\hat{\mathbf{f}}$ is a minimizer of a cost function $J(\mathbf{f})$ of the form

$$J(\mathbf{f}) = \Theta(H\mathbf{f} - \mathbf{g}) + \beta\Phi(\mathbf{f}),$$

where Θ forces closeness to data according to (7.1), Φ embodies the priors, and $\beta > 0$ is a parameter that controls the trade-off between these two terms. The most usual choice for Θ is

$$\Theta(\mathbf{v}) = \|\mathbf{v}\|_2^2. \tag{7.2}$$

In a statistical setting, this Θ corresponds to an assumption that $\mathbf{n}$ is the white Gaussian noise. Our computational method is designed for Θ as given in (7.2) or any smooth and convex function. Recently, data terms of the form $\Theta(\mathbf{v}) = \|\mathbf{v}\|_1$ were shown to be useful if some data entries have to be preserved, which is appreciable, for instance, if $\mathbf{n}$ is impulse noise [9, 20, 28]. Our method is straightforward to extend to this situation; see for instance [17].

We now consider the choice of Φ. In many image processing applications, the regularization term Φ reads

$$\Phi(\mathbf{f}) = \sum_{i \in G} \varphi(\|\mathrm{D}_i\mathbf{f}\|_2), \tag{7.3}$$

where $G = \{1, \ldots, p\}$ is the set of all pixels of the image, $\mathrm{D}_i : \mathbb{R}^p \to \mathbb{R}^s$, for $s \geq 1$ and $i \in G$, is a linear operator yielding a vector containing the differences between pixel i and its neighbors, and φ is called a potential function. Each D_i can be seen as an $s \times p$ matrix. When $\varphi(t) = |t|$ and $\{\mathrm{D}_i\}$ corresponds to the discrete analogy of the gradient operator, the function (7.3) is the well-known total variation (TV) regularization function [35].

Various potential functions have been used in the literature; a complete review can be found in [5]. An important requirement is that the potential function φ allows the recovery of large differences $\|\mathrm{D}_i\mathbf{f}\|_2$ at the right locations of edges and smoothes the other differences. It is well known that this requirement cannot be met by $\varphi(t) = t^2$ which was originally used in [38]. Since the pioneering work [14] by Geman and Geman, different nonconvex functions φ have been considered either in a statistical or in a variational framework [1, 4, 12, 13, 15, 22, 24, 33]. In order to avoid the numerical intricacies arising with nonconvex regularization, an important effort was done to derive *convex* edge-preserving potential functions [6, 10, 16, 21, 35, 37]. Nevertheless, *nonconvex nonsmooth regularization* offers much richer possibilities to restore high quality images with neat edges: a theoretical explanation was provided in [29, 30], while numerical examples can be found in numerous articles [12, 13, 27, 34, 31].

In this chapter, we derive fast algorithms to approximate faithfully the global minimizer of these nonconvex and nonsmooth cost functions in section 7.2. In section 7.3, our experimental results show clearly the effectiveness and efficiency of the proposed numerical schemes. Concluding remarks and perspectives are sketched in section 7.4.

7.2 Minimization methods

The minimization of nonconvex function J involves three major difficulties which drastically restrict the methods that can be envisaged. Because of the nonconvexity of φ, the function J may exhibit a large number of local minima which are not global. In addition, the function J is usually nonsmooth at the minimizers, and thus usual gradient-based methods are inappropriate for local minimization. Finally, the matrix H can have numerous nonzero elements beyond the diagonal and is often ill-conditioned. Global minimization of J can be considered using either stochastic algorithms or continuation-based deterministic relaxation.

In [31], a nonsmooth graduated nonconvexity (GNC) continuation method is proposed to solve the nonconvex nonsmooth minimization problem. Consider a sequence

$$\varepsilon_0 = 0 < \varepsilon_1 < \cdots < \varepsilon_k < \cdots < \varepsilon_l = 1.$$

We define that φ is approximated by a sequence φ_{ε_k} such that φ_0 is convex and $\varphi_1 = \varphi$. As ε_k increases from 0 to 1, the nonconvexity of φ_{ε_k} is gradually increased. It has been shown by experiments that the GNC method can provide better minimization results for the nonconvex nonsmooth objective functionals and is less sensitive to the initial guess compared with the non-continuation methods in [31]. To simplify the notations, we write ε for ε_k whenever this is clear from the context. Therefore the function J can be approximated by a sequence of function J_ε correspondingly:

$$J_\varepsilon(\mathbf{f}) = \|H\mathbf{f} - \mathbf{g}\|_2^2 + \beta \sum_{i \in G} \varphi_\varepsilon(\|\mathrm{D}_i \mathbf{f}\|_2). \tag{7.4}$$

We split φ_ε into two terms. The first term ψ_ε is smooth but nonconvex. The other term $|t|$ is convex but nonsmooth, which is a TV type regularization, i.e., $\varphi_\varepsilon(t) = \psi_\varepsilon(t) + \alpha_\varepsilon |t|$, where $\psi_\varepsilon(t) = \varphi_\varepsilon(t) - \varphi_\varepsilon{}'(0^+)|t|$ and $\alpha_\varepsilon = \varphi_\varepsilon{}'(0^+)$. Then $J_\varepsilon(\mathbf{f})$ in (7.4) can be further rewritten as

$$J_\varepsilon(\mathbf{f}) = \|H\mathbf{f} - \mathbf{g}\|_2^2 + \beta \sum_{i \in G} \psi_\varepsilon(\|\mathrm{D}_i \mathbf{f}\|_2) + \beta \alpha_\varepsilon \sum_{i \in G} \|\mathrm{D}_i \mathbf{f}\|_2, \tag{7.5}$$

which contains a data fidelity term with ℓ_2 norm, a nonconvex term, and a nonsmooth TV regularization term. The main aim of this chapter is to use the nonsmooth GNC continuation method for solving (7.5). We remark that $\psi_\varepsilon(\|\mathrm{D}_i \mathbf{f}\|_2)$ and $\|\mathrm{D}_i \mathbf{f}\|_2$ in [31] are approximated by $\psi_\varepsilon(\mathbf{d}_i^T \mathbf{f})$ and $|\mathbf{d}_i^T \mathbf{f}|$ respectively, where $\mathbf{d}_i^T : \mathbb{R}^p \to \mathbb{R}$, for $i \in G$, are linear operators. Usually they are the first-order differences between each pixel and its 4 or 8 adjacent neighbors. In such case, the problem can be reformulated using linear programming techniques. More precisely, this nonconvex nonsmooth minimization problem is given as the minimization of a nonconvex differentiable objective function under a set of linear constraints and it can be solved by the modified primal-dual interior point method, which guarantees the descent direction at each step. Experimental results showed that this method can provide better performance compared to a simulated annealing method with significantly less computational cost. In this chapter, the use of the ℓ_2 norm of the discrete gradient, i.e., $\|\mathrm{D}_i \mathbf{f}\|_2$ instead of $|\mathbf{d}_i^T \mathbf{f}|$, yields image restorations of better quality.

Now, we propose our algorithm. Our idea is to tackle the difficulty for minimizing the function which consists of a nonconvex term and a nonsmooth term, where we use the variable-splitting and the penalty technique to separate them with respect to two different variables. Based on the idea of [18], an auxiliary variable $\mathbf{u}$ is used to transfer the nonsmooth

TV term out of $\mathbf{f}$. A fitting term is then added to J_ε to ensure the closeness of $\mathbf{f}$ and $\mathbf{u}$. Therefore, we have

$$J_\varepsilon(\mathbf{f},\mathbf{u}) = \|H\mathbf{f}-\mathbf{g}\|_2^2 + \beta\sum_{i\in G}\psi_\varepsilon(\|\mathrm{D}_i\mathbf{f}\|_2) + \omega\|\mathbf{f}-\mathbf{u}\|_2^2 + \beta\alpha_\varepsilon\sum_{i\in G}\|\mathrm{D}_i\mathbf{u}\|_2,$$

where ω is the regularization parameter for the fitting term $\|\mathbf{f}-\mathbf{u}\|_2^2$. By fixing the variable $\mathbf{u}$, J_ε is a smooth function with respect to $\mathbf{f}$. Therefore, it can be minimized by gradient-based methods. By fixing the variable $\mathbf{f}$, minimizing J_ε equals to the TV denoising problem which can be solved by the existing method efficiently [7]. We will demonstrate in section 7.3 that the computational cost of the proposed method is greatly reduced compared with that of the interior point method proposed in [31], while the qualities of restored images are competitive by the experimental results.

Let us state the algorithm in detail. For each ε, starting from an initial guess $\mathbf{u}^{(0)}$, this method computes a sequence of iterates:

$$\mathbf{f}^{(1)}, \mathbf{u}^{(1)}, \mathbf{f}^{(2)}, \mathbf{u}^{(2)}, \ldots, \mathbf{f}^{(j)}, \mathbf{u}^{(j)}, \ldots,$$

where

$$\mathbf{f}^{(j)} = \arg\min_{\mathbf{f}} J_\varepsilon^h(\mathbf{f},\mathbf{u}^{(j-1)})$$

with

$$J_\varepsilon^h(\mathbf{f},\mathbf{u}) = \|H\mathbf{f}-\mathbf{g}\|_2^2 + \beta\sum_{i\in G}\psi_\varepsilon(\|\mathrm{D}_i\mathbf{f}\|_2) + \omega\|\mathbf{f}-\mathbf{u}\|_2^2,$$

and

$$\mathbf{u}^{(j)} = \arg\min_{\mathbf{u}} J_\varepsilon^{tv}(\mathbf{f}^{(j)},\mathbf{u})$$

with

$$J_\varepsilon^{tv}(\mathbf{f},\mathbf{u}) = \omega\|\mathbf{f}-\mathbf{u}\|_2^2 + \beta\alpha_\varepsilon TV(\mathbf{u}).$$

The first step of this algorithm is to minimize J_ε^h. For $\varepsilon = 0$, we have

$$\sum_{i\in G}\psi_\varepsilon(\|\mathrm{D}_i\mathbf{f}\|_2) = 0, \ \forall\mathbf{f}.$$

The minimization of J_ε^h is exactly the same as the TV deblurring problem in [18]. Therefore, we can follow the framework in [18] to minimize J_ε^h. For $\varepsilon > 0$, since all terms in J_ε^h are differentiable, we can find the gradient vector and the Hessian of J_ε^h to tackle the minimization problem:

$$\nabla_{\mathbf{f}} J_\varepsilon^h = 2H^T(H\mathbf{f}-\mathbf{g}) + \beta\nabla_{\mathbf{f}}\sum_{i\in G}\psi_\varepsilon(\|\mathrm{D}_i\mathbf{f}\|_2) + 2\omega(\mathbf{f}-\mathbf{u})$$

and

$$\nabla_{\mathbf{f}}^2 J_\varepsilon^h = 2H^TH + 2\omega I + \beta\nabla_{\mathbf{f}}^2\sum_{i\in G}\psi_\varepsilon(\|\mathrm{D}_i\mathbf{f}\|)_2,$$

where I is the identity matrix. For the Hessian matrix in the quasi-Newton method, $\nabla_{\mathbf{f}}^2 J_\varepsilon^h$ may not be positive definite since it contains a negative definite term $\nabla_{\mathbf{f}}^2 \sum_{i\in G}\psi_\varepsilon(\|\mathrm{D}_i\mathbf{f}\|)_2$. In order to ensure the convergence of the algorithm, we simply use the positive definite part of the Hessian matrix in the optimization procedure. Because of the term $2\omega I$, the coefficient matrix $2H^TH + 2\omega I$ is always positive definite even if H^TH is singular. The quasi-Newton method can then be used to solve J_ε^h [32]. We remark that in image restoration H is usually a blurring matrix generated by a symmetric point spread function. The computational cost of the quasi-Newton method is dominated by three fast transforms for performing matrix-vector multiplication with H [26]. The computational cost for each fast transform is only $O(p \log p)$ for a $p \times p$ blurring matrix [26]. After finding the descent direction $\Delta \mathbf{f}^{(j-1)}$ by solving

$$(2H^TH + 2\omega I)\Delta \mathbf{f}^{(j-1)} = -\nabla_{\mathbf{f}} J_\varepsilon^h,$$

the step size τ of the method is then determined by line search method in order to find $\mathbf{f}^{(j)} = \mathbf{f}^{(j-1)} + \tau \Delta \mathbf{f}^{(j-1)}$. Three different conditions have been considered, including Armijo rule, Goldstein rule, and a fixed τ [32]. However, the experimental results suggest us to use a fixed τ as it requires less computational cost, compared with the other two line search methods. The second step of this method is to apply an exact TV denoising problem scheme to $\mathbf{f}^{(j)}$. Since the function J_ε^{tv} is identically the same as that in [18], we employ the Chambolle projection algorithm [7] to solve this problem. For ε_k, we simply choose a linearly increasing ε_k in our experiments as suggested in [31].

7.3 Experimental results

In this section, we present several experimental results to demonstrate the efficiency of the proposed method. Signal to noise ratio (SNR) is used to measure the quality of the restored images, while CPU time is also used to compare the efficiency of the restoration method. The stopping criterion of our proposed method is that the relative difference between the restored images of successive iterations should satisfy the following inequality:

$$\frac{\|\mathbf{f}^{(j+1)} - \mathbf{f}^{(j)}\|_2}{\|\mathbf{f}^{(j+1)}\|_2} < 10^{-4}.$$

The potential function used in all the illustrations is:

$$\varphi(t) = \frac{\alpha|t|}{1+\alpha|t|}, \qquad \varphi_\varepsilon(t) = \frac{\alpha|t|}{1+\varepsilon\alpha|t|}, \quad 0 \le \varepsilon \le 1,$$

which has been tested in [31]. We compare the proposed method to the modified interior point method in [31]. The conjugate gradient method is used to solve the inner linear systems arising from outer iterations, and the stopping criterion of the conjugate gradient method is that the residual should be less than 10^{-5}. We remark that the proposed method does not require to solve any inner linear systems.

Four images are tested, which are all gray images with intensity values ranging from 0 to 1. The first image is an artificial circles image of size 64×64, and the second one is cameraman of size 256×256. These two images have been tested in [31]. The third image is tank image of size 512×512. To generate the observed images, we added the Gaussian noise with a standard deviation of 0.05 with blurring. The blurring function is chosen to be a two dimensional Gaussian function

$$h(i,j) = e^{-2(i/3)^2 - 2(j/3)^2},$$

which is truncated such that the function has a support of 7×7. The fourth image is the modified Shepp-Logan image of size 50×50. To generate the observed image for this image,

Table 7.1: Restored SNRs and CPU times for different images.

	The interior point method		The proposed method	
Image	Computational time	SNR	Computational time	SNR
Circle	32.89	19.85	9.91	19.43
Cameraman	743.11	22.87	34.38	23.47
Tank	3015.09	31.17	40.02	31.31

we added the Gaussian noise with a standard deviation of 0.05. The Radon transform is used to construct the degradation matrix H. These images are further transformed by back-projection such that H can be reformulated as a convolution operator.

Different initial guesses have been considered, including the observed image, the least squares solution, and a flat image (all pixel values are 0.5). From our experimental results, both methods are insensitive to all of the initial guesses. Therefore, we only demonstrate the results in which the initial guesses are the observed images. According to the potential function, we can set $\alpha_\varepsilon = 0.5$. We tested different values of β and ω in order to find the restored image with the highest SNR among the tested values. Similarly, we also tested different values of the regularization parameter in the modified interior point method to find the restored image with the highest SNR.

7.3.1 Test of blurred images

Figures 7.1–7.3(a) show the original images. Figures 7.1–7.3(b) show their corresponding images with blur and noise as described in the above settings respectively. Figures 7.1–7.3(c) and 7.1–7.3(d) show the images restored by the modified interior point method and the proposed method respectively. For simplicity, we use the same set of parameters for restoring images in Figures 7.1–7.3(c) and 7.1–7.3(d). We see from the figures that the images restored by the modified interior point method and the proposed method are visually almost the same. In Table 7.1, we show their SNR results, and find that they are almost the same. However, the computational time (in seconds) required by the proposed method is significantly lower than that by the interior point method. These results demonstrate the proposed method is quite efficient for restoring images.

7.3.2 Test of Radon transform images

In this subsection, the reconstruction of the images transformed by the Radon transform using our proposed method is presented. The modified Shepp-Logan image is applied to illustrate the efficiency of our algorithm. Following the example in [31], we set the image to be of size 50×50. Figure 7.4(a) is the original modified Shepp-Logan image. The image is transformed along the angles from 0^0 to 180^0 with the increasing of 6 degrees to create the size of 75×31 Radon transform of the original image. The noise from normal distribution with mean zero and standard deviation 0.05 is added to this transformed image to generate the observed image in Figure 7.4(b). The degradation matrix H in this example is the discrete Radon transform matrix which cannot be reformulated as a convolution operator. In order to restore the image and maintain the efficiency of the proposed method, we make use of the back-projection operator B [19]. It can be shown that the back-projected Radon

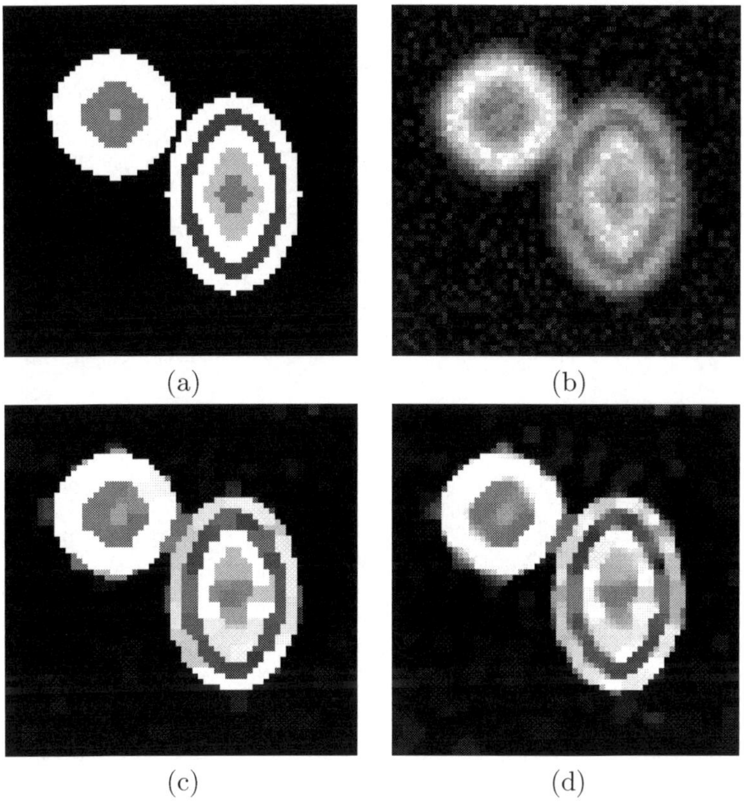

Figure 7.1: (a) The original circle image; (b) the observed image; (c) the image restored by the interior point method with $\beta = 0.015$; (d) the image restored by the proposed method with $\omega = 5.0$ and $\beta = 0.015$.

Figure 7.2: (a) The original cameraman image; (b) the observed image; (c) the image restored by the interior point method with $\beta = 0.03$; (d) the image restored by the proposed method with $\omega = 7.0$ and $\beta = 0.03$.

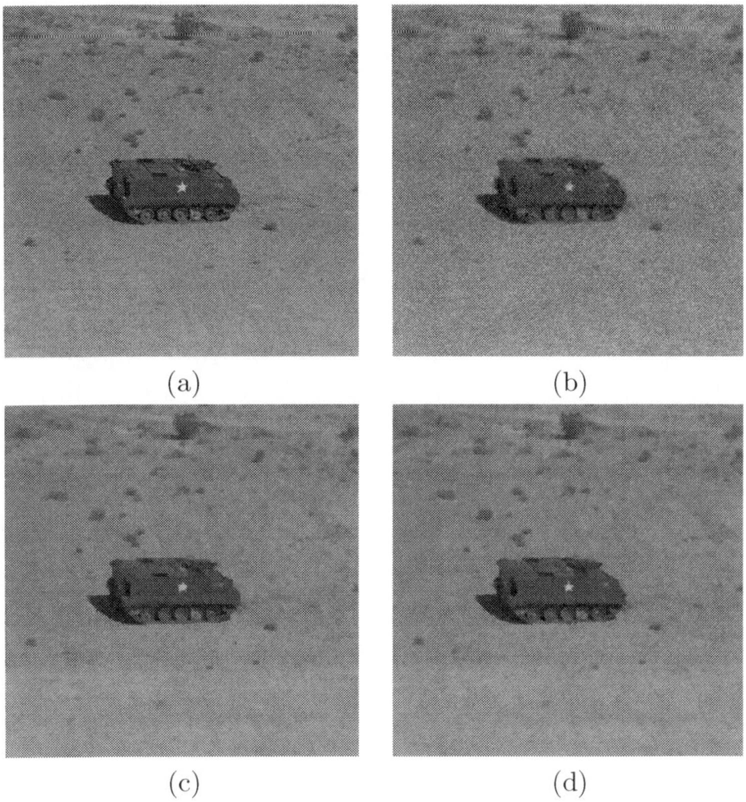

(a) (b) (c) (d)

Figure 7.3: (a) The original tank image; (b) the observed image; (c) the image restored by the interior point method with $\beta = 0.07$; (d) the image restored by the proposed method with $\omega = 4.0$ and $\beta = 0.07$.

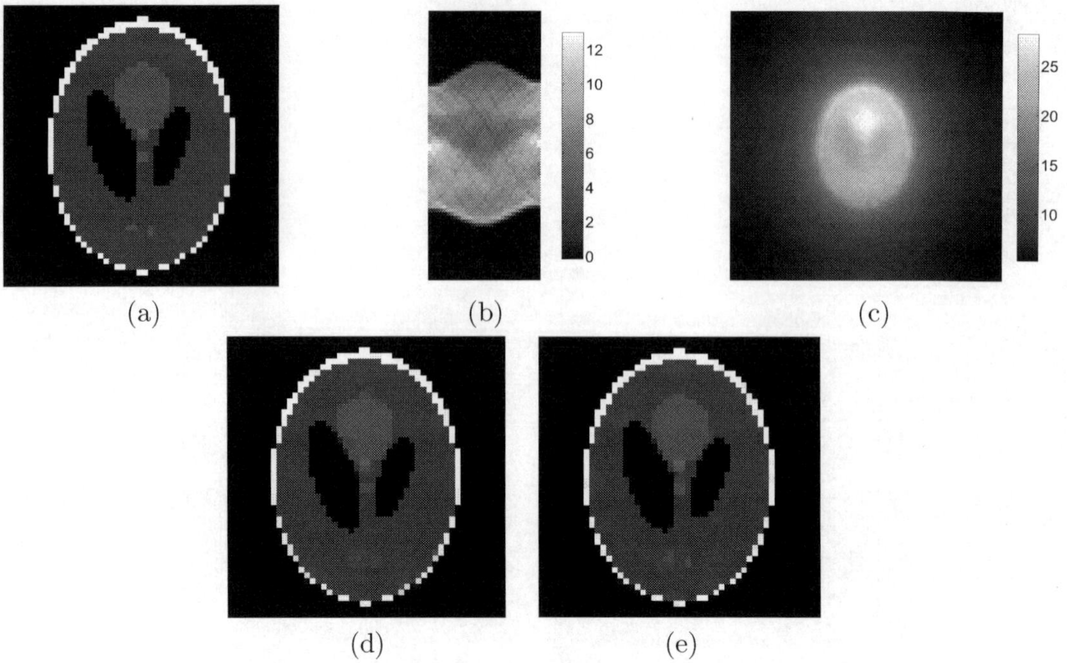

Figure 7.4: (a) The original modified Shepp-Logan image with size 50×50; (b) the obtained image after the Radon transform along the angles from 0^0 to 180^0 with the increasing of 6 degrees; (c) the back-projected Radon transform image; (d) the image restored from Figure 7.4(b) by the interior point method with $\beta = 0.8$; (e) the image restored from Figure 7.4(c) by the proposed method with $\omega = 0.5$ and $\beta = 0.04$.

transform is an image of $\mathbf{f}$ blurred by the point spread function of the form $(x^2 + y^2)^{-\frac{1}{2}}$ which can be used to construct the convolution matrix H. Now we solve

$$J_\varepsilon^h(\mathbf{f}, \mathbf{u}) = \|H\mathbf{f} - B\mathbf{g}\|_2^2 + \beta \sum_{i \in G} \psi_\varepsilon(\|\mathrm{D}_i \mathbf{f}\|_2) + \omega \|\mathbf{f} - \mathbf{u}\|_2^2.$$

The back-projected Radon transform $B\mathbf{g}$ is shown in Figure 7.4(c). Figure 7.4(d) shows the resulting image that is reconstructed from Figure 7.4(b) by the interior point method. Figure 7.4(e) shows the resulting image reconstructed from Figure 7.4(c) by the proposed method. Both of the methods provide high quality restored images. However, as conjugate gradient method is required to solve the linear system, the interior point method took more time (1860 seconds) to reconstruct the image. The proposed method only takes 2 seconds to restore about the same quality of the reconstructed image. The SNRs of the reconstructed image by the interior point method and the proposed method are 41.96dB and 43.68dB.

7.4 Concluding remarks

In this chapter, we considered the image reconstruction and restoration by the nonconvex and nonsmooth regularization in ℓ_2 norm of the discrete gradient of the image. The main goal is to conceive a fast numerical scheme to solve the difficult minimization problem. We used a nonsmooth GNC method to replace the global minimization of the nonconvex nonsmooth functional by a family of local minimizations. We then applied the variable-splitting and penalty technique to separate the nonconvex and the nonsmooth terms with respect to two different variables. By fixing one variable and minimizing another variable, the minimization problem was split into a TV deblurring problem and a TV denoising problem. The first problem was solved by the Quasi-Newton method and the second problem was solved by Chambolle projection algorithm. The total computational cost is dominated by three fast transforms with Chambolle projection method for each iteration. Numerical experiments have been given to show the effectiveness and efficiency of the proposed algorithm.

References for Chapter 7

[1] G. Aubert and P. Kornprobst, *Mathematical Problems in Image Processing*, Springer-Verlag, Berlin, 2nd ed., 2006.

[2] J. F. Aujol, G. Gilboa, T. Chan, and S. Osher, *Structure-texture image decomposition–modeling, algorithms, and parameter selection*, International Journal of Computer Vision, 67 (2006), pp. 111–136.

[3] M. R. Banham and A. K. Katsaggelos, *Digital image restoration*, IEEE Signal Processing Magazine, 14 (1997), pp. 24–41.

[4] J. E. Besag, *Digital image processing: towards Bayesian image analysis*, J. Appl. Stat., 16 (1989), pp. 395–407.

[5] M. Black and A. Rangarajan, *On the unification of line processes, outlier rejection, and robust statistics with applications to early vision*, International Journal of Computer Vision, 19 (1996), pp. 57–91.

[6] C. Bouman and K. Sauer, *A generalized Gaussian image model for edge-preserving MAP estimation*, IEEE Trans. Image Processing, 2 (1993), pp. 296–310.

[7] A. Chambolle, *An algorithm for total variation minimization and applications*, Journal of Mathematical Imaging and Vision, 20 (2004), pp. 89–97.

[8] C. L. Chan, A. K. Katsaggelos, and A. V. Sahakian, *Image sequence filtering in quantum-limited noise with applications to low-dose fluoroscopy*, IEEE Trans. Medical Imaging, 12 (1993), pp. 610–621.

[9] T. Chan and S. Esedoglu, *Aspects of total variation regularized ℓ_1 function approximation*, Technical Report, University of California at Los Angeles, CAM report, 2004.

[10] P. Charbonnier, L. Blanc-Féraud, G. Aubert, and M. Barlaud, *Deterministic edge-preserving regularization in computed imaging*, IEEE Trans. Image Processing, 6 (1997), pp. 298–311.

[11] G. Demoment, *Image reconstruction and restoration: overview of common estimation structure and problems*, IEEE Trans. Acoustics Speech and Signal Processing, ASSP-37 (1989), pp. 2024–2036.

[12] D. Geman and G. Reynolds, *Constrained restoration and recovery of discontinuities*, IEEE Trans. Pattern Anal. Machine Intell., PAMI-14 (1992), pp. 367–383.

[13] D. Geman and C. Yang, *Nonlinear image recovery with half-quadratic regularization*, IEEE Trans. Image Processing, 4 (1995), pp. 932–946.

[14] S. Geman and D. Geman, *Stochastic relaxation, Gibbs distributinons, and the Bayesian restoration of images*, IEEE Trans. Pattern Anal. Machine Intell., PAMI-6 (1984), pp. 721–741.

[15] S. Geman and D. E. McClure, *Statistical methods for tomographic image reconstruction*, Proc. of the 46th Session of the ISI, Bulletin of the ISI, 52 (1987), pp. 22–26.

[16] P. J. Green, *Bayesian reconstructions from emission tomography data using a modified EM algorithm*, IEEE Trans. Medical Imaging, 9 (1990), pp. 84–93.

[17] X. Guo, F. Li, and M. Ng, *A fast $\ell 1$-TV algorithm for image restoration*, SIAM Journal on Scientific Computing, 31 (2009), pp. 2322–2341.

[18] Y. Huang, M. Ng, and Y. Wen, *A fast total variation minimization method for image restoration*, SIAM Journal on Multiscale Modeling and Simulation, 7 (2008), pp. 774–795.

[19] A. Jain, *Fundamentals of Digital Image Processing*, Prentice-Hall, Englewood Cliffs, NJ, 1989.

[20] T. Kärkkäinen, K. Kunisch, and K. Majava, *Denoising of smooth images using l^1-fitting*, Computing, 74 (2005), pp. 353–376.

[21] K. Lange, *Convergence of EM image reconstruction algorithms with Gibbs priors*, IEEE Trans. Medical Imaging, 9 (1990), pp. 439–446.

[22] S. Li, *Markov Random Field Modeling in Computer Vision*, Springer-Verlag, New York, 1st ed., 1995.

[23] Z. K. Liu and J. Y. Xiao, *Restoration of blurred TV pictures caused by uniform linear motion*, Comput. Vision, Graphics, Image Proc., 44 (1988), pp. 30–34.

[24] D. Mumford and J. Shah, *Boundary detection by minimizing functionals*, Proceedings of the IEEE Int. Conf. on Computer Version and Pattern Recognition, San Francisco, June 1985, pp. 22–26.

[25] M. R. Nagel, *Introduction to evaluation of motion-degraded images*, Proc. of NASA Electronics Research Center Seminar, Cambridge, MA, Dec. 1968, pp. 3–5.

[26] M. Ng, R. Chan, and W. Tang, *A fast algorithm for deblurring models with Neumann boundary conditions*, SIAM Journal on Scientific Computing, 21 (1999), pp. 851–866.

[27] M. Nikolova, *Markovian reconstruction using a GNC approach*, IEEE Trans. Image Processing, 8 (1999), pp. 1204–1220.

[28] M. Nikolova, *A variational approach to remove outliers and impulse noise*, Journal of Mathematical Imaging and Vision, 20 (2004), pp. 99–120.

[29] M. Nikolova, *Analysis of the recovery of edges in images and signals by minimizing nonconvex regularized least-squares*, SIAM Journal on Multiscale Modeling and Simulation, 4 (2005), pp. 960–991.

[30] M. Nikolova, *Analytical bounds on the minimizers of (nonconvex) regularized least-squares*, AIMS Journal on Inverse Problems and Imaging, 4 (2007), pp. 661–677.

[31] M. Nikolova, M. Ng, S. Zhang, and W. Ching, *Efficient reconstruction of piecewise constant images using nonsmooth nonconvex minimization*, SIAM Journal on Imaging Sciences, 1 (2008), pp. 2–25.

[32] J. Nocedal and S. Wright, *Numerical Optimization*, Springer-Verlag, New York, 1999.

[33] P. Perona and J. Malik, *Scale-space and edge detection using anisotropic diffusion*, IEEE Trans. Pattern Anal. Machine Intell., PAMI-12 (1990), pp. 629–639.

[34] M. Robini, A. Lachal, and I. Magnin, *A stochastic continuation approach to piecewise constant reconstruction*, IEEE Trans. Image Processing, 16 (2007), pp. 2576–2589.

[35] L. Rudin, S. Osher, and C. Fatemi, *Nonlinear total variaion based noise removal algorithm*, Physica D: Nonlinear Phenomena, 60 (1992), pp. 259–268.

[36] C. H. Slump, *Real-time image restoration in diagnostic X-ray imaging, the effects on quantum noise*, Proc, 11th IAPR Int. Conf. on Pattern Recognition, vol. II, Conference B: Pattern Recognition Methodology and Systems, 1992, pp. 693–696.

[37] R. Stevenson and E. Delp, *Fitting curves with discontinuities*, Proc. of the 1st Int. Workshop on Robust Comput. Vision, Seattle, WA, 1990, pp. 127–136.

[38] A. Tikhonov and V. Arsenin, *Solutions of Ill-Posed Problems*, Winston, Washington DC, 1977.

[39] Y. Wang, J. Yang, W. Yin, and Y. Zhang, *A new alternating minimization algorithm for total variation image reconstruction*, SIAM Journal on Imaging Sciences, 1 (2008), pp. 248–272.

Chapter 8

Eigenvalues of Certain Augmented Complex Stochastic Matrices with Applications to PageRank

Gang Wu [1]

Abstract

Given a row stochastic matrix P of size n, we examine relationships between the eigenvalues of P and those of some augmented matrices with respect to P. Applications of the results to Google's PageRank problem are also discussed.

8.1 Introduction

Google is one of the most sucessful search engine in recent years, most due to its very comprehensive and accurate search results. An important technology, PageRank, plays an important role in Google's search engine. Nowadays, PageRank has emerged as a popular link analysis model, mostly due to its query-independence, using only the Web graph structure [8].

Let us consider the hyperlink structure of the Web as a directed graph. Suppose the Web of interest contains n pages, then the Markov chain model represents this graph with an n-by-n transition probability matrix G. Each element of G is the probability of moving from node i to node j in one time step. However, G may not be stochastic. This happens whenever a node contains no outlinks, which is called dangling node. One remedy is to replace all zero rows with $\frac{1}{n}\mathbf{e}^T$, where $\mathbf{e} = (1, 1, \ldots, 1)^T$ is the column vector of all ones.

[1]School of Mathematical Sciences, Xuzhou Normal University, Xuzhou, 221116, Jiangsu, P. R. China (email: gangwu76@yahoo.com.cn or wugangzy@gmail.com).
The author is supported by the National Science Foundation of China under grant 10901132, the Natural Science Foundation for Colleges and Universities in Jiangsu Province under grant 08KJB110012, the Qing-Lan Project of Jiangsu Province under grant QL200612, and the Natural Science Foundation of Xuzhou Normal University under grant 08XLY01.

Then we obtain a row stochastic matrix

$$P = G + \mathbf{d}\mathbf{e}^T/n,$$

where $\mathbf{d} = (d_1, d_2, \ldots, d_n)^T$ is a column vector of zeros and ones. An element d_i is equal to 1 if the ith row of G corresponding to a dangling node, and is 0 otherwise. Unfortunately, this adjustment matrix alone is not enough to insure the existence of the stationary distribution of the Markov chain, i.e., the existence of the PageRank vector. When the chain is irreducible, the PageRank vector is guaranteed to exist [1].

Indeed, the Markov chain matrix P produced from the hyperlink structure of the Web is almost certainly reducible. Thus a positive long-run stationary vector will not exist for the subsequent Markov chain [1, 6]. The original solution of Brin and Page makes use of the method of *the maximal irreducibility*, whereby every node is directly connected to every other node, hence irreducibility is trivially enforced [8]. The famous Google matrix is defined as a convex combination of P and a rank-one matrix

$$A = \alpha P + (1-\alpha)\mathbf{e}\mathbf{v}^T = \alpha G + \frac{\alpha}{n}\mathbf{d}\mathbf{e}^T + (1-\alpha)\mathbf{e}\mathbf{v}^T,$$

where $\mathbf{v}$ is a nonnegative vector whose elements sum to one, called personalization vector; and $0 < \alpha < 1$ is the damping factor. The PageRank vector is noting but the (left) dominant eigenvector of the Google matrix [6]. The relationship between the spectrum (the set of all the eigenvalues) of P and that of A has been well understood [4, 6]. However, the relationship between the spectrum of P and that of the transition matrix G is still unknown [6].

Now let us discuss this topic briefly. Denote the spectrum of G by $\Lambda(G) = \{\lambda_1, \lambda_2, \ldots, \lambda_n\}$ and that of P by $\Lambda(P) = \{\mu_1, \mu_2, \ldots, \mu_n\}$. Set $\Gamma = \Lambda(G) \cup \Lambda(P)$. We can separate Γ into three disconnected sets

$$\Gamma = \Lambda_1 \cup \Lambda_2 \cup \Lambda_3,$$

where

$$\Lambda_1 = \{\lambda \in \Gamma \mid \lambda \in \Lambda(G) \text{ and } \lambda \in \Lambda(P)\},$$

$$\Lambda_2 = \{\lambda \in \Gamma \mid \lambda \in \Lambda(G) \text{ and } \lambda \notin \Lambda(P)\},$$

and

$$\Lambda_3 = \{\lambda \in \Gamma \mid \lambda \in \Lambda(P) \text{ and } \lambda \notin \Lambda(G)\}.$$

For any $\lambda \in \Lambda_2$, we have

$$\det(G - \lambda I) = \det(P - \lambda I) \cdot \det\big(I - (P - \lambda I)^{-1}\mathbf{d}\mathbf{e}^T/n\big) = 0,$$

where I is the identity matrix, which implies

$$\det\big(I - (P - \lambda I)^{-1}\mathbf{d}\mathbf{e}^T/n\big) = 1 - \mathbf{e}^T(P - \lambda I)^{-1}\mathbf{d}/n = 0,$$

because $\det(\lambda I - P) \neq 0$. That is, if $\lambda \in \Lambda_2$, then

$$\mathbf{e}^T(P - \lambda I)^{-1}\mathbf{d} = n.$$

Similarly, we have for any $\lambda \in \Lambda_3$,

$$\det\big(I + (G - \lambda I)^{-1}\mathbf{d}\mathbf{e}^T/n\big) = 1 + \mathbf{e}^T(G - \lambda I)^{-1}\mathbf{d}/n = 0.$$

That is, if $\lambda \in \Lambda_3$, then

$$\mathbf{e}^T(G-\lambda I)^{-1}\mathbf{d} = -n.$$

However, the maximal irreducibility does alter the Web's true nature, whereas other methods of forcing irreducibility seem less invasive and more inline with the Web's true nature [6]. One alternative is the method of *the minimal irreducibility* [9]. In this type of method, a dummy node is added to the Web, which connects to every other node and to which every other node is connected, making the chain irreducible in a minimal sense. One way of creating a minimally irreducible Markov matrix is [6]:

$$\hat{P} = \begin{bmatrix} \alpha P & (1-\alpha)\mathbf{e} \\ \mathbf{v}^T & 0 \end{bmatrix} \in \mathbb{R}^{(n+1)\times(n+1)}, \tag{8.1}$$

where $0 < \alpha < 1$ is the damping factor and $\mathbf{v}$ is a probability vector called the personalization vector. The following theorem compares the spectrum of P and that of $\hat{P}$.

Theorem 8.1 **[6]** *Given a stochastic matrix P with spectrum $\{1, \lambda_2, \ldots, \lambda_n\}$, the spectrum of $\hat{P}$ is $\{1, \alpha\lambda_2, \ldots, \alpha\lambda_n, \alpha - 1\}$.*

Another way to create a minimally irreducible Markov matrix is [7]:

$$\breve{P} = \begin{bmatrix} \frac{n}{n+1}P & \frac{1}{n+1}\mathbf{e} \\ \frac{1}{n+1}\mathbf{e}^T & \frac{1}{n+1} \end{bmatrix} \in \mathbb{R}^{(n+1)\times(n+1)}. \tag{8.2}$$

The relationship between the spectrum of P and that of $\breve{P}$ is described in the following theorem.

Theorem 8.2 **[7]** *Given a stochastic matrix P with spectrum $\{1, \lambda_2, \ldots, \lambda_n\}$, the spectrum of $\breve{P}$ is $\left\{1, \frac{n}{n+1}\lambda_2, \ldots, \frac{n}{n+1}\lambda_n, 0\right\}$.*

Indeed, both $\hat{P}$ and $\breve{P}$ can be viewed as augmented stochastic matrices with respect to P. More precisely, $\hat{P}$ and $\breve{P}$ are obtained by augmenting αP with one row and one column, and the two matrices are still stochastic. In this chapter, we generalize (8.1) and (8.2) to their block cases, i.e., we augment αP with k rows and k columns

$$\overline{\hat{P}} = \begin{bmatrix} \alpha P & \alpha_1\mathbf{e} & \cdots & \alpha_k\mathbf{e} \\ \mathbf{v}_1^* & & & \\ \vdots & & \mathbf{O} & \\ \mathbf{v}_k^* & & & \end{bmatrix} \text{ and } \overline{\breve{P}} = \begin{bmatrix} \alpha P & \gamma\mathbf{e} & \cdots & \gamma\mathbf{e} \\ \bar{\gamma}\mathbf{e}^T & & & \\ \vdots & & \beta\cdot\mathbf{L} & \\ \bar{\gamma}\mathbf{e}^T & & & \end{bmatrix}, \tag{8.3}$$

under the condition that $\overline{\hat{P}}\mathbf{e} = \overline{\breve{P}}\mathbf{e} = \mathbf{e}$. Here $\mathbf{O}$ is the k-by-k zero matrix, $\mathbf{L}$ is the k-by-k one matrix whose elements are all equal to 1, $\mathbf{v}_i^*$ denotes the conjugate transpose of $\mathbf{v}_i$, and $\bar{\gamma}$ represents the conjugate of γ. We call $\overline{\hat{P}}$ and $\overline{\breve{P}}$ *augmented matrices* with respect to P. These two models relate to adding k (more than one) dummy nodes to the Web, which connects to every other node and to which every other node is connected, making the chain irreducible in a minimal sense. Therefore, (8.3) can be viewed as a natural generalization of (8.1) and (8.2), where only one dummy node is added.

Given the spectrum of P, finding the spectra of the augmented matrices is an interesting work. In [6, 7], Langville and Meyer considered the spectra of (8.1) and (8.2) respectively. In this chapter, we generalize the results of Theorems 8.1 and 8.2 to their block cases, and give insight into the eigenvalues of $\overline{\hat{P}}$ and $\overline{\breve{P}}$ respectively. Furthermore, the new results are more general in the sense that the matrices involved can be complex.

Throughout this chapter, we denote by P an n-by-n row stochastic matrix, by $\mathbf{e}$ the vector of all ones with its dimension clear from the context, by $\mathbf{O}$ the k-by-k zero matrix, and by $\mathbf{L}$ the k-by-k one matrix. Let I and O be the identity matrix and the zero matrix whose size are clear from the context.

8.2 Main results

Given a row stochastic matrix P with spectrum $\{1, \lambda_2, \ldots, \lambda_n\}$, we are interested in the spectrum of the following matrix

$$\overline{\hat{P}} = \begin{bmatrix} \alpha P & \alpha_1 \mathbf{e} & \cdots & \alpha_k \mathbf{e} \\ \mathbf{v}_1^* & & & \\ \vdots & & \mathbf{O} & \\ \mathbf{v}_k^* & & & \end{bmatrix} \in \mathbb{C}^{(n+k)\times(n+k)},$$

where $\alpha_1, \ldots, \alpha_k \in \mathbb{C}$ with $\sum_{i=1}^{k} \alpha_i = 1 - \alpha$, and $\mathbf{v}_1, \ldots, \mathbf{v}_k \in \mathbb{C}^n$ with $\mathbf{v}_i^* \mathbf{e} = 1$ for $1 \le i \le k$. We first propose a useful lemma.

Lemma 8.1 *Let the spectrum of P be $\{1, \lambda_2, \ldots, \lambda_n\}$ and $\tilde{P} = \gamma P + [\delta_1 \mathbf{e}, \ldots, \delta_k \mathbf{e}] W^*$, where $\gamma, \delta_1, \ldots, \delta_k \in \mathbb{C}$ and $W = [\mathbf{w}_1, \mathbf{w}_2, \ldots, \mathbf{w}_k] \in \mathbb{C}^{n\times k}$. Then the spectrum of $\tilde{P}$ is*

$$\Big\{\gamma + \sum_{i=1}^{k} \delta_i (\mathbf{w}_i^* \mathbf{e}), \gamma\lambda_2, \ldots, \gamma\lambda_n\Big\}.$$

Proof: Let $P = URU^*$ be the Schur decomposition of P, where R is upper triangular with diagonal elements $\{1, \lambda_2, \ldots, \lambda_n\}$ and $U = [\mathbf{e}/\|\mathbf{e}\|_2,\ U_\perp]$ is unitary. Then

$$U^* \tilde{P} U = \gamma R + U^* [\delta_1 \mathbf{e}, \ldots, \delta_k \mathbf{e}] W^* U. \tag{8.4}$$

Note that

$$U^* [\delta_1 \mathbf{e}, \ldots, \delta_k \mathbf{e}] = \begin{bmatrix} \frac{\mathbf{e}^T}{\|\mathbf{e}\|_2} \\ U_\perp^* \end{bmatrix} [\delta_1 \mathbf{e}, \ldots, \delta_k \mathbf{e}] = \begin{bmatrix} \delta_1 \|\mathbf{e}\|_2, \ldots, \delta_k \|\mathbf{e}\|_2 \\ O \end{bmatrix},$$

where we used $U_\perp^* \mathbf{e} = O$. Moreover,

$$W^* U = \begin{bmatrix} \mathbf{w}_1^* \\ \vdots \\ \mathbf{w}_k^* \end{bmatrix} \Big[\frac{\mathbf{e}}{\|\mathbf{e}\|_2}, U_\perp\Big] = \begin{bmatrix} \frac{\mathbf{w}_1^* \mathbf{e}}{\|\mathbf{e}\|_2} & \\ \vdots & W^* U_\perp \\ \frac{\mathbf{w}_k^* \mathbf{e}}{\|\mathbf{e}\|_2} & \end{bmatrix},$$

where $W^*U_\perp \in \mathbb{C}^{k\times(n-1)}$. Therefore,

$$U^*\left[\delta_1\mathbf{e},\ldots,\delta_k\mathbf{e}\right]W^*U = \begin{bmatrix} \delta_1\|\mathbf{e}\|_2,\ldots,\delta_k\|\mathbf{e}\|_2 \\ O \end{bmatrix}\cdot\begin{bmatrix} \frac{\mathbf{w}_1^*\mathbf{e}}{\|\mathbf{e}\|_2} & \\ \vdots & W^*U_\perp \\ \frac{\mathbf{w}_k^*\mathbf{e}}{\|\mathbf{e}\|_2} & \end{bmatrix}$$

$$= \begin{bmatrix} \sum_{i=1}^{k}\delta_i(\mathbf{w}_i^*\mathbf{e}) & Z \\ O & O \end{bmatrix}, \tag{8.5}$$

where $Z = \left[\delta_1\|\mathbf{e}\|_2,\ldots,\delta_k\|\mathbf{e}\|_2\right]\cdot W^*U_\perp$ is a row vector of dimension $n-1$. Therefore, we obtain from (8.4) and (8.5) that $U^*\tilde{P}U$ is upper triangular with diagonal elements

$$\left\{\gamma+\sum_{i=1}^{k}\delta_i(\mathbf{w}_i^*\mathbf{e}),\gamma\lambda_2,\ldots,\gamma\lambda_n\right\},$$

which is the spectrum of $\tilde{P}$. □

Remark 8.1 *Recall that the columns of* $\left[\delta_1\mathbf{e},\ldots,\delta_k\mathbf{e}\right]$ *are not the eigenvectors of* P *corresponding to* $1,\lambda_2,\ldots,\lambda_k$. *Therefore, the results presented in* [2, 3, 5, 10, 11] *do not apply here.*

The following theorem gives a relationship between the spectrum of P and that of $\overline{\hat{P}}$, which generalizes the result obtained in Theorem 8.1.

Theorem 8.3 *Let the spectrum of* P *be* $\{1,\lambda_2,\ldots,\lambda_n\}$. *Denote by*

$$\overline{\hat{P}} = \begin{bmatrix} \alpha P & S \\ V^* & \mathbf{O} \end{bmatrix} \in \mathbb{C}^{(n+k)\times(n+k)},$$

where $\alpha\in\mathbb{C}$, $S=\left[\alpha_1\mathbf{e},\ldots,\alpha_k\mathbf{e}\right]\in\mathbb{C}^{n\times k}$ *with* $\sum_{i=1}^{k}\alpha_i = 1-\alpha$, *and* $V=\left[\mathbf{v}_1,\ldots,\mathbf{v}_k\right]\in\mathbb{C}^{n\times k}$ *with* $\mathbf{v}_i^*\mathbf{e}=1$ *for* $1\le i\le k$. *Then the spectrum of* $\overline{\hat{P}}$ *is*

$$\{1,\alpha\lambda_2,\ldots,\alpha\lambda_n,\alpha-1,0,\ldots,0\}.$$

Proof: We prove the theorem in three steps.

(i) Firstly, we show that

$$\Lambda(\overline{\hat{P}}) = \Lambda(\alpha P - XV^*)\cup\Lambda(V^*X),$$

where $X = \left[\frac{\alpha_1}{1-\alpha}\mathbf{e},\frac{\alpha_2}{1-\alpha}\mathbf{e},\ldots,\frac{\alpha_k}{1-\alpha}\mathbf{e}\right] = \frac{1}{1-\alpha}S$ and $\Lambda(M)$ denotes the spectrum of a matrix M.

Let $Q = \begin{bmatrix} I & X \\ O & I \end{bmatrix}$ and then $Q^{-1} = \begin{bmatrix} I & -X \\ O & I \end{bmatrix}$. We have

$$Q^{-1}\overline{\hat{P}}Q = \begin{bmatrix} \alpha P - XV^* & \alpha PX + S - XV^*X \\ V^* & V^*X \end{bmatrix}. \tag{8.6}$$

It is easy to verify that

$$\alpha PX = \frac{\alpha}{1-\alpha}[\alpha_1\mathbf{e},\ldots,\alpha_k\mathbf{e}], \qquad XV^*X = \frac{1}{1-\alpha}[\alpha_1\mathbf{e},\ldots,\alpha_k\mathbf{e}],$$

where we used $P\mathbf{e} = \mathbf{e}$, $\mathbf{v}_i^*\mathbf{e} = 1$ for $1 \le i \le k$, and $\sum_{i=1}^{k} \alpha_i = 1-\alpha$. Therefore,

$$\alpha PX + S - XV^*X = O. \tag{8.7}$$

We have by (8.6) and (8.7),

$$Q^{-1}\overline{\hat{P}}Q = \begin{bmatrix} \alpha P - XV^* & O \\ V^* & V^*X \end{bmatrix}.$$

Obviously,

$$\Lambda(\overline{\hat{P}}) = \Lambda(\alpha P - XV^*) \cup \Lambda(V^*X).$$

(ii) Secondly, it follows from Lemma 8.1 that

$$\Lambda(\alpha P - XV^*) = \{\alpha - 1, \alpha\lambda_2, \ldots, \alpha\lambda_n\}.$$

(iii) Finally, we show that $\Lambda(V^*X) = \{1, 0, \ldots, 0\}$.
Recall that

$$V^*X = \begin{bmatrix} \frac{\alpha_1}{1-\alpha} & \cdots & \frac{\alpha_k}{1-\alpha} \\ \vdots & & \vdots \\ \frac{\alpha_1}{1-\alpha} & \cdots & \frac{\alpha_k}{1-\alpha} \end{bmatrix} = \frac{1}{1-\alpha}\mathbf{e}\mathbf{g}^*,$$

where $\mathbf{g}^* = (\alpha_1, \ldots, \alpha_k)$. Then we have from [12] that

$$\Lambda(V^*X) = \left\{\frac{1}{1-\alpha}\mathbf{g}^*\mathbf{e}, 0, \ldots, 0\right\} = \{1, 0, \ldots, 0\}.$$

We thus complete the proof of the theorem by combining (i), (ii), and (iii). □

Now we focus on the matrix of the form

$$\overline{\breve{P}} = \begin{bmatrix} \alpha P & \gamma\mathbf{e} & \cdots & \gamma\mathbf{e} \\ \bar{\gamma}\mathbf{e}^T & & & \\ \vdots & & \beta \cdot \mathbf{L} & \\ \bar{\gamma}\mathbf{e}^T & & & \end{bmatrix} \in \mathbb{C}^{(n+k)\times(n+k)},$$

where $\alpha \in \mathbb{C}$, $\gamma = \frac{1-\alpha}{k}$, and $\beta = \frac{k-(1-\bar{\alpha})n}{k^2}$.

The following theorem establishes a relationship between the spectrum of P and that of $\overline{\breve{P}}$, which generalizes the result of Theorem 8.2.

Theorem 8.4 *Let the spectrum of P be $\{1, \lambda_2, \ldots, \lambda_n\}$. Denote by*

$$\overline{\breve{P}} = \begin{bmatrix} \alpha P & \breve{X} \\ \breve{X}^* & Y \end{bmatrix} \in \mathbb{C}^{(n+k)\times(n+k)},$$

where $\alpha \in \mathbb{C}$, $\breve{X} = \gamma[\mathbf{e},\ldots,\mathbf{e}] \in \mathbb{C}^{n\times k}$, $Y = \beta \cdot \mathbf{L}$, $\gamma = \frac{1-\alpha}{k}$, and $\beta = \frac{k-(1-\bar{\alpha})n}{k^2}$. Then the spectrum of $\overline{\breve{P}}$ is

$$\left\{1, \alpha\lambda_2, \ldots, \alpha\lambda_n, \alpha - \frac{1-\bar{\alpha}}{k}n, 0, \ldots, 0\right\}.$$

Proof: Again, we prove the theorem in three steps.

(i) Firstly, we prove that

$$\Lambda(\overline{\breve{P}}) = \Lambda(\alpha P - \tilde{X}\breve{X}^*) \cup \Lambda(\breve{X}^*\tilde{X} + Y),$$

where $\tilde{X} = \frac{1}{k}[\mathbf{e},\ldots,\mathbf{e}] \equiv \frac{1}{k}E$, with $E = [\mathbf{e},\ldots,\mathbf{e}] \in \mathbb{R}^{n\times k}$. Let $Q = \begin{bmatrix} I & \tilde{X} \\ O & I \end{bmatrix}$ and then $Q^{-1} = \begin{bmatrix} I & -\tilde{X} \\ O & I \end{bmatrix}$. We have

$$Q^{-1}\overline{\breve{P}}Q = \begin{bmatrix} \alpha P - \tilde{X}\breve{X}^* & \alpha P\tilde{X} + \breve{X} - \tilde{X}\breve{X}^*\tilde{X} - \tilde{X}Y \\ \breve{X}^* & \breve{X}^*\tilde{X} + Y \end{bmatrix}. \tag{8.8}$$

It is easy to see that

$$\alpha P\tilde{X} = \frac{\alpha}{k}[\mathbf{e},\ldots,\mathbf{e}] = \frac{\alpha}{k}E, \quad \breve{X}^*\tilde{X} = \frac{\bar{\gamma}}{k}E^TE = \frac{\bar{\gamma}n}{k}\mathbf{L}, \quad \tilde{X}\breve{X}^*\tilde{X} = \frac{1}{k}E\cdot\frac{\bar{\gamma}n}{k}\mathbf{L} = \frac{\bar{\gamma}n}{k}E,$$

where we used $E^TE = n\cdot\mathbf{L}$ and $E\cdot\mathbf{L} = kE$. Furthermore,

$$\tilde{X}Y = \frac{1}{k}E\cdot\beta\mathbf{L} = \beta E.$$

Note that $\frac{\alpha}{k} + \gamma - \frac{\bar{\gamma}n}{k} - \beta = 0$ and consequently,

$$\alpha P\tilde{X} + \breve{X} - \tilde{X}\breve{X}^*\tilde{X} - \tilde{X}Y = O. \tag{8.9}$$

We have from (8.8) and (8.9),

$$\Lambda(\overline{\breve{P}}) = \Lambda(\alpha P - \tilde{X}\breve{X}^*) \cup \Lambda(\breve{X}^*\tilde{X} + Y).$$

(ii) Secondly, it follows from Lemma 8.1 that

$$\Lambda(\alpha P - \tilde{X}\breve{X}^*) = \left\{\alpha - \frac{1-\bar{\alpha}}{k}n, \alpha\lambda_2, \ldots, \alpha\lambda_n\right\}.$$

(iii) Finally, we prove that $\Lambda(\breve{X}^*\tilde{X} + Y) = \{1, 0, \ldots, 0\}$. Indeed,

$$\breve{X}^*\tilde{X} + Y = \frac{\bar{\gamma}n}{k}\mathbf{L} + \beta\cdot\mathbf{L} = \frac{(1-\bar{\alpha})n}{k^2}\mathbf{L} + \left[\frac{1}{k} - \frac{(1-\bar{\alpha})n}{k^2}\right]\mathbf{L} = \frac{1}{k}\cdot\mathbf{L}.$$

Note that the eigenvalues of $\frac{1}{k}\mathbf{L}$ are $\{1, 0, \ldots, 0\}$. Thus $\Lambda(\breve{X}^*\tilde{X} + Y) = \{1, 0, \ldots, 0\}$. We therefore complete the proof of the theorem by combining (i), (ii), and (iii). □

Remark 8.2 *When $\overline{\hat{P}}$ and $\overline{\breve{P}}$ are nonnegative such that they are stochastic matrices, the above results still hold. Furthermore, it can be seen that* Theorems 8.3 *and* 8.4 *are more general in the sense that the matrices involved can be complex.*

Acknowledgements I would like to express my sincere thanks to Professor Xiao-qing Jin for inviting me to submit a paper to the Proceedings. Moreover, the author is very grateful to Professor Yi-min Wei for helpful discussions.

References for Chapter 8

[1] A. Berman and R. Plemmons, *Nonnegative Matrices in the Mathematical Sciences*, SIAM, Philadelphia, 1994.

[2] J. Ding and G. Yao, *The eigenvalue problem of a specifically updated matrix*, Appl. Math. Comput., 185 (2007), pp. 415–420.

[3] J. Ding and A. Zhou, *A spectral theorem for rank-one updated matrices with some applications*, Appl. Math. Letters, 20 (2007), pp. 1223–1226.

[4] L. Eldén, *The eigenvalues of the Google matrix*, Technical Report, LiTH-MAT-RC2004-01.

[5] R. Horn and S. Serra-Capizzano, *A general setting for the parametric Google matrix*, Internet Math., 3 (2006), pp. 385–411.

[6] A. Langville and C. Meyer, *Deeper inside PageRank*, Internet Math., 1 (2004), pp. 335–380.

[7] A. Langville and C. Meyer, *Fiddling with PageRank*, Techniqal Report, Department of Mathematics, North Carolina State University, 2003.

[8] L. Page, S. Brin, R. Motwami, and T. Winograd, *The PageRank citation ranking: bring order to the Web*, Technical report, Computer Science Department, Stanford University, 1998.

[9] J. Tomlin, *A new paradigm for ranking pages on the World Wide Web*, The 12th International World Wide Web Conference, 2003.

[10] G. Wu, *Eigenvalues and Jordan canonical form of a successively rank-one updated complex matrix with applications to Google's PageRank problem*, J. Comput. Appl. Math., 216 (2008), pp. 364–370.

[11] G. Wu and Y. Wei, *Comments on "Jordan canonical form of the Google matrix"*, SIAM J. Matrix Anal. Appl., 30 (2008), pp. 364–374.

[12] F. Zhang, *Matrix Theory: Basic Results and Techniques*, Springer, 1999.

Chapter 9

Clenshaw-Curtis-Rational Quadrature Rule for Wiener-Hopf Equations of the Second Kind

Yan Xuan [1] **Fu-rong Lin** [2]

Abstract

This chapter considers numerical solutions for Wiener-Hopf equations of the second kind defined on the half-line $[0, \infty)$:

$$y(t) + \int_0^\infty K(t-s)y(s)\mathrm{d}s = g(t), \qquad 0 \le t < \infty.$$

We first combine the well-known Clenshaw-Curtis quadrature with a rational variable transformation to obtain a Clenshaw-Curtis-rational (CCR) quadrature rule, which is suitable for integrals on semi-infinite intervals. We then apply the CCR quadrature to Wiener-Hopf equations. The reduction of singularities in Wiener-Hopf equations is also considered. Numerical examples are given to illustrate the efficiency of the CCR quadrature.

9.1 Introduction

Wiener-Hopf equation of the second kind is of the form

$$y(t) + \int_0^\infty K(t-s)y(s)\mathrm{d}s = g(t), \qquad 0 \le t < \infty, \tag{9.1}$$

where $K(t) \in L_1(\mathbb{R})$ and $g(t) \in L_p[0, \infty)$ $(1 \le p < \infty)$ are given functions. It arises in various applications in mathematics and engineering; see [9, pp. 145–146 and pp. 186–189] for instance.

[1]Department of Mathematics, Shantou University, Shantou Guangdong 515063, China (email: yxaun1@stu.edu.cn).

[2]Department of Mathematics, Shantou University, Shantou Guangdong 515063, China (email: frlin@stu.edu.cn).

The approximate solution of such an equation has attracted a lot of attention. Many authors considered truncating the half-line integral equation to obtain a finite-section one

$$y_\tau(t) + \int_0^\tau K(t-s)y_\tau(s)\mathrm{d}s = g(t), \qquad 0 \le t \le \tau;$$

see for instance [1, 2, 6, 7, 10, 11, 12, 14, 15]. It has been proved [6] that

$$\lim_{\tau\to\infty} \|y_\tau - y\|_{L_p[0,\tau]} = 0, \quad 1 \le p < \infty,$$

for $g(t) \in L_p[0,\infty)$. In 1990s, the preconditioned conjugate gradient method for finite-section Wiener-Hopf equations was introduced by Gohberg et al. [7]. They used circulant integral operators defined by periodic kernel functions as preconditioners. Later on, Lin et al. [12] proposed to use the convolution operator as a preconditioner instead of the circulant operator. In 2000s, Kang et al. proposed a highly accurate numerical scheme for finite-section Wiener-Hopf integral equations based on the Clenshaw-Curtis quadrature [10, 11].

There are studies on numerical solutions for the half-line equations without truncation [3, 8, 13]. In [3], the authors applied Nyström methods based on composite quadratures to the equation. Under the assumption that the exact solution decays exponentially, they proved that by properly choosing quadrature points, the numerical solution converges to the exact solution. Under the same assumption, Mastroianni and Monegato [13] developed a numerical method based on a product quadrature using the zeros of Laguerre polynomials. In [8], Graham and Mendes considered the case where the kernel function has logarithmic singularity at $s = 0$ and decays exponentially as $s \to \infty$.

In this chapter, we also consider solving the equation without truncation. We first derive a numerical quadrature for $\int_0^\infty f(s)ds$ in section 9.2. By combining the Clenshaw-Curtis quadrature with the rational variable transformation $s = \frac{\alpha(1-z)}{z+1}$ ($\alpha > 0$ is a parameter), we obtain the Clenshaw-Curtis-rational (CCR) quadrature rule:

$$\int_0^\infty f(s)ds \approx \sum_{j=1}^n \frac{2\alpha w_j}{(z_j+1)^2} f\left(\frac{\alpha(1-z_j)}{z_j+1}\right),$$

where z_j and w_j, $j = 1, 2, \ldots, n$, are the quadrature points and weights of the n-point Clenshaw-Curtis quadrature (to be specified in section 9.2). We then analyze the accuracy of the CCR quadrature.

In section 9.3, we discuss the application of the CCR quadrature to (9.1). To reduce the singularities, we reformulate (9.1) as

$$\left[1 + \int_0^\infty K(t-s)ds\right]y(t) + \int_0^\infty K(t-s)[y(s) - y(t)]ds = g(t), \quad 0 \le t < \infty. \tag{9.2}$$

Applying the CCR quadrature to $\int_0^\infty K(t-s)[y(s) - y(t)]ds$, we get the following linear system

$$\left[1 + \int_0^\infty K(t_i - s)ds - \sum_{j=1}^n K(t_i - t_j)\frac{2\alpha w_j}{(z_j+1)^2}\right]y_i + \sum_{j=1}^n K(t_i - t_j)\frac{2\alpha w_j}{(z_j+1)^2}y_j = g(t_i),$$

for $i = 1, 2, \ldots, n$, where $t_i = \frac{\alpha(1-z_i)}{z_i+1}$ and y_i are approximate values of $y(t_i)$. The above reformulation is called subtraction of singularity [5, pp. 268–270]. In section 9.3, we also introduce another method to reduce the singularities.

The outline of this chapter is as follows. In section 9.2, we derive a CCR quadrature rule and analyze its accuracy. In section 9.3, we apply the CCR quadrature to Wiener-Hopf equations, discuss the singularities in the equations, and introduce two methods to reduce them. Numerical examples with several types of kernel functions are given in section 9.4. Finally, concluding remarks are given in section 9.5.

9.2 CCR quadrature rule

In this section, we derive a CCR quadrature rule and analyze its accuracy.

9.2.1 The quadrature rule

Clenshaw-Curtis quadrature rules [4] are closely related to Chebyshev polynomials. Let

$$T_0(z) = 1, \quad T_i(z) = \cos(i \arccos z), \quad i = 1, 2, \ldots,$$

be the sequence of Chebyshev polynomials. The quadrature rule with quadrature points

$$z_k = \cos\left(\frac{(2k-1)\pi}{2n}\right), \quad k = 1, 2, \ldots, n, \tag{9.3}$$

the roots of $T_n(z)$, is called the classical Clenshaw-Curtis quadrature, and the one with quadrature points $\bar{z}_k = \cos(k\pi/n)$, the roots of the polynomial $[T_{n+1}(z) - T_{n-1}(z)]$, is called the practical Clenshaw-Curtis quadrature. In the following, we only introduce the classical Clenshaw-Curtis quadrature. We refer to [16] for a practical one.

There are many ways to derive the Clenshaw-Curtis quadrature rule. Here we introduce the quadrature in a way similar to the one presented in [5]. Let $f(z) \in L_2[-1, 1]$, we have the Chebyshev-Fourier expansion:

$$f(z) = \frac{b_0}{2} + \sum_{i=1}^{\infty} b_i T_i(z), \tag{9.4}$$

where

$$b_i = \frac{2}{\pi} \int_{-1}^{1} \frac{f(z)T_i(z)}{\sqrt{1-z^2}} \mathrm{d}z, \quad i = 0, 1, 2, \ldots. \tag{9.5}$$

We obtain from (9.4),

$$I(f) \equiv \int_{-1}^{1} f(z)\mathrm{d}z = b_0 + \sum_{i=1}^{\infty} b_i \int_{-1}^{1} T_i(z)\mathrm{d}z = b_0 + \sum_{i=1}^{\infty} \frac{2b_{2i}}{1-4i^2}. \tag{9.6}$$

Applying the n-point Gauss-Chebyshev quadrature to (9.5), we get

$$b_i \approx \tilde{b}_i = \frac{2}{n} \sum_{k=1}^{n} f(z_k)T_i(z_k) = \frac{2}{n} \sum_{k=1}^{n} f(z_k) \cos\left(\frac{i(2k-1)\pi}{2n}\right), \quad i = 0, 1, 2, \ldots. \tag{9.7}$$

Recall that the n-point Gauss-Chebyshev quadrature is given by

$$\int_{-1}^{1} \frac{g(z)}{\sqrt{1-z^2}} \mathrm{d}z \approx \frac{\pi}{n} \sum_{k=1}^{n} g(z_k), \tag{9.8}$$

where $z_k = \cos\left(\frac{(2k-1)\pi}{2n}\right)$, $k = 1, 2, \ldots, n$, are given by (9.3). Combining (9.6) with (9.7), we get the Clenshaw-Curtis quadrature rule

$$\begin{aligned} Q_n(f) &\equiv \tilde{b}_0 + \sum_{i=1}^{n-1} \tilde{b}_i \int_{-1}^{1} T_i(z)dz = \tilde{b}_0 + \sum_{i=1}^{\lfloor (n-1)/2 \rfloor} \frac{2\tilde{b}_{2i}}{1-4i^2} \\ &= \frac{1}{n} \sum_{k=1}^{n} \left[2f(z_k) + \sum_{i=1}^{\lfloor (n-1)/2 \rfloor} \frac{4}{1-4i^2} \cos\left(\frac{i(2k-1)\pi}{n}\right) f(z_k) \right] \\ &= \sum_{k=1}^{n} w_k f(z_k), \end{aligned} \tag{9.9}$$

where $\lfloor x \rfloor$ denotes the largest integer that is not larger than x and

$$w_k = \frac{1}{n}\left[2 + \sum_{i=1}^{\lfloor (n-1)/2 \rfloor} \frac{4}{1-4i^2} \cos\left(\frac{i(2k-1)\pi}{n}\right) \right], \quad k = 1, \ldots, n. \tag{9.10}$$

We note that the quadrature points and weights of the Clenshaw-Curtis quadrature (9.9) are given by (9.3) and (9.10) respectively. Furthermore, the weights can be efficiently computed via fast Fourier transform (FFT) techniques in $O(n \log n)$ operations.

Let $\mathcal{P}_k$ denotes the set of all polynomials of degree being less than or equal to k. Note that for $g(z) \in \mathcal{P}_{2n-1}$, the n-point Gauss-Chebyshev quadrature (9.8) exactly holds. It follows that for $f(z) \in \mathcal{P}_{n-1}$, $\tilde{b}_i = b_i$, $i = 0, 1, \ldots, n-1$. Therefore, when $f(z) \in \mathcal{P}_{n-1}$, the n-point Clenshaw-Curtis quadrature (9.9) gives the exact value, i.e., $Q_n(f) = I(f)$. Recall that the n-point Gauss-Legendre quadrature is more accurate since it gives the exact integral for functions in $\mathcal{P}_{2n-1}$. However, for the n-point Gauss-Legendre quadrature, the computation of quadrature points and weights requires $O(n^3)$ operations.

To apply the Clenshaw-Curtis quadrature to $\int_0^\infty f(s)ds$, we introduce a new variable

$$z = \frac{\alpha(1-s)}{s+\alpha}, \quad \text{i.e.,} \quad s = \frac{\alpha(1-z)}{z+1},$$

where $\alpha > 0$ is a parameter. Then

$$\int_0^\infty f(s)ds = \int_{-1}^{1} f\left(\frac{\alpha(1-z)}{z+1}\right) \frac{2\alpha}{(z+1)^2} dz.$$

If we apply the Gauss-Legendre quadrature to the right-hand side of the above integral, we get the Gauss-rational quadrature [5]. Applying the n-point Clenshaw-Curtis quadrature to the above integral, we have

$$\int_0^\infty f(s)ds \approx \sum_{j=1}^{n} \frac{2\alpha w_j}{(z_j+1)^2} f\left(\frac{\alpha(1-z_j)}{z_j+1}\right) \equiv Q_n^{(CCR)}(f), \tag{9.11}$$

where z_j, $j = 1, 2, \ldots, n$, are the roots of $T_n(z)$ given by (9.3), and w_j, $j = 1, 2 \ldots, n$, are given by (9.10). We call (9.11) a Clenshaw-Curtis-rational (CCR) quadrature rule.

9.2.2 Accuracy estimate

In this subsection, we analyze the accuracy of the Clenshaw-Curtis quadrature (9.9) and the CCR quadrature (9.11). The following lemmas are about the coefficients b_i $(i = 0, 1, \ldots)$ and the approximate coefficients $\tilde{b}_i$ $(i = 0, 1, \ldots, n-1)$ given by (9.5) and (9.7) respectively.

Lemma 9.1 ([11]) *Let $f \in C^r[-1, 1]$ with $r > 1$ as*

$$f(z) = \frac{b_0}{2} + \sum_{i=1}^{\infty} b_i T_i(z), \quad -1 \le z \le 1.$$

Then there exists a constant $c > 0$ such that

$$|b_i| \le \frac{c}{i^r} \quad \text{and} \quad |f(z) - f_n(z)| \le \frac{c}{r-1}\frac{1}{n^{r-1}},$$

where $f_n(z) = b_0/2 + \sum_{i=1}^{n-1} b_i T_i(z)$.

Lemma 9.2 *Let $f(z) = \frac{b_0}{2} + \sum_{i=1}^{\infty} b_i T_i(z)$ and $\tilde{b}_i$, $i = 0, 1, \ldots, n-1$, be given by (9.7). Then*

$$\tilde{b}_i = b_i + \sum_{l=1}^{\infty} (-1)^l (b_{2nl+i} + b_{2nl-i}), \quad i = 0, 1, \ldots, n-1. \tag{9.12}$$

Proof: From (9.7) we see that

$$\tilde{b}_0 = \frac{2}{n}\sum_{k=1}^{n}\left[\frac{b_0}{2} + \sum_{j=1}^{\infty} b_j T_j(z_k)\right] = b_0 + \frac{2}{n}\sum_{j=1}^{\infty} b_j \sum_{k=1}^{n} T_j(z_k).$$

Let

$$S_j = \sum_{k=1}^{n} T_j(z_k) = \sum_{k=1}^{n} \cos\left(\frac{j(2k-1)\pi}{2n}\right) = \sum_{k=1}^{n} \cos((2k-1)\theta_j),$$

where $\theta_j = \frac{j\pi}{2n}$. Notice that for j satisfying $2n \nmid j$,

$$\sum_{k=1}^{n} \cos\left((2k-1)\theta_j\right) = \frac{\sin(2n\theta_j)}{2\sin\theta_j} = \frac{\sin(j\pi)}{2\sin\theta_j} = 0,$$

it is easily seen that

$$S_j = \begin{cases} 0, & 2n \nmid j, \\ (-1)^l n, & j = 2nl. \end{cases} \tag{9.13}$$

Therefore

$$\tilde{b}_0 = b_0 - 2b_{2n} + 2b_{4n} - \cdots = b_0 + \sum_{l=1}^{\infty} (-1)^l (b_{2nl+0} + b_{2nl-0}).$$

Note that for $i = 1, 2, \ldots, n-1$,

$$\begin{aligned}
\tilde{b}_i &= \frac{2}{n}\sum_{k=1}^{n}\Big[\frac{b_0}{2} + \sum_{j=1}^{\infty} b_j T_j(z_k)\Big] T_i(z_k) \\
&= \frac{b_0}{n}\sum_{k=1}^{n} T_i(z_k) + \frac{2}{n}\sum_{j=1}^{\infty} b_j \sum_{k=1}^{n} T_j(z_k) T_i(z_k) \\
&= \frac{b_0}{n}\sum_{k=1}^{n} T_i(z_k) + \frac{1}{n}\sum_{j=1}^{\infty} b_j \Big[\sum_{k=1}^{n} T_{j-i}(z_k) + \sum_{k=1}^{n} T_{j+i}(z_k)\Big] \\
&= \frac{b_0}{n} S_i + \frac{1}{n}\sum_{j=1}^{\infty} b_j S_{j-i} + \frac{1}{n}\sum_{j=1}^{\infty} b_j S_{j+i}.
\end{aligned}$$

We obtain by using (9.13),

$$\tilde{b}_i = \sum_{l=0}^{\infty}(-1)^l b_{2nl+i} + \sum_{l=1}^{\infty}(-1)^l b_{2nl-i} = b_i + \sum_{l=1}^{\infty}(-1)^l (b_{2nl+i} + b_{2nl-i}).$$

Thus, (9.12) follows. □

The following theorem gives the accuracy of the Clenshaw-Curtis quadrature.

Theorem 9.1 *Let $f \in C^r[-1,1]$ with $r > 1$ and $Q_n(f)$ be the Clenshaw-Curtis quadrature defined by* (9.9). *Then*

$$|I(f) - Q_n(f)| = \left|\int_{-1}^{1} f(z)dz - \sum_{j=1}^{n} w_j f(z_j)\right| = O(n^{-r}).$$

Proof: Since $f(z) \in C^r[-1,1]$, by Lemma 9.1, there exists a constant $c > 0$ such that

$$|b_i| \le c i^{-r}.$$

Assume that n is even. From (9.6) and (9.9), we have

$$\begin{aligned}
|I(f) - Q_n(f)| &= \left| b_0 + \sum_{i=1}^{\infty} \frac{2b_{2i}}{1-4i^2} - \tilde{b}_0 - \sum_{i=1}^{n/2-1} \frac{2\tilde{b}_{2i}}{1-4i^2} \right| \\
&\le |b_0 - \tilde{b}_0| + \sum_{i=1}^{n/2-1} \frac{2|b_{2i} - \tilde{b}_{2i}|}{4i^2-1} + \sum_{i=n/2}^{\infty} \frac{2|b_{2i}|}{4i^2-1}. \qquad (9.14)
\end{aligned}$$

We obtain by using (9.12),

$$|b_0 - \tilde{b}_0| = \left|2\sum_{l=1}^{\infty}(-1)^{l-1} b_{2nl}\right| \le 2c\sum_{l=1}^{\infty}(2nl)^{-r} = 2^{-r+1} c n^{-r}\sum_{l=1}^{\infty} l^{-r} = c_1 n^{-r},$$

and, for $i = 1, 2, \ldots, n/2 - 1$,

$$|b_{2i} - \tilde{b}_{2i}| \le \sum_{l=1}^{\infty}(|b_{2nl+2i}| + |b_{2nl-2i}|) \le c\sum_{l=1}^{\infty}\big[(2nl+2i)^{-r} + (2nl-2i)^{-r}\big] \le c_2 n^{-r},$$

where c_1 and c_2 are constants. Therefore

$$\begin{aligned}|b_0 - \tilde{b}_0| + \sum_{i=1}^{n/2-1} \frac{2|b_{2i} - \tilde{b}_{2i}|}{4i^2 - 1} &\leq \max\{c_1, c_2\} n^{-r} \Big[1 + \sum_{i=1}^{n/2-1} \frac{2}{(2i-1)(2i+1)}\Big] \\ &\leq \max\{c_1, c_2\} n^{-r} \Big(2 - \frac{1}{n-1}\Big) = O(n^{-r}). \qquad (9.15)\end{aligned}$$

Moreover,

$$\sum_{i=n/2}^{\infty} \frac{2|b_{2i}|}{4i^2-1} \leq c \sum_{i=n/2}^{\infty} \frac{2(2i)^{-r}}{4i^2-1} < cn^{-r} \sum_{i=n/2}^{\infty} \frac{2}{(2i-1)(2i+1)} = O(n^{-r-1}). \qquad (9.16)$$

From (9.14), (9.15), and (9.16), the theorem follows. □

From the proof of Theorem 9.1, we see that the accuracy of the Clenshaw-Curtis quadrature is determined by the decay rate of the coefficients in the expansion of $f(z)$. We emphasize that $f \in C^r[-1,1]$ is not a necessary condition. For example, let $f(z) = (1+z)^{3/2}$. We have $f(z) \in C^1[-1,1]$ and $f(z) \notin C^2[-1,1]$. However, the expansion coefficients of $f(z)$ are given by

$$b_i = \frac{48\sqrt{2}}{\pi(4i^2-9)(4i^2-1)} = O(i^{-4}).$$

Therefore, when the Clenshaw-Curtis quadrature is used to evaluate $\int_{-1}^{1}(1+z)^{3/2}dz$, the accuracy is $O(n^{-4})$ rather than $O(n^{-1})$.

In the following, we analyze the accuracy of the CCR quadrature. We first prove Lemma 9.3 that is required for proving the main result in Theorem 9.2.

Lemma 9.3 *Let $r \geq 2$ be an integer. Then*

$$\frac{1}{2} - \frac{1}{3}\binom{r-2}{1} + \cdots + (-1)^{r-1}\frac{1}{r-1}\binom{r-2}{r-3} + (-1)^r\frac{1}{r} = \frac{1}{r(r-1)}, \qquad (9.17)$$

where $\displaystyle\binom{m}{n} = \frac{m!}{n!(m-n)!}$.

Proof: Consider the polynomial

$$p(z) = \frac{z^2}{2} - \frac{z^3}{3}\binom{r-2}{1} + \cdots + (-1)^{r-1}\frac{z^{r-1}}{r-1}\binom{r-2}{r-3} + (-1)^r\frac{z^r}{r}.$$

We only require to prove that $p(1) = \frac{1}{r(r-1)}$. The derivative of $p(z)$ is

$$\begin{aligned}p'(z) &= z - z^2\binom{r-2}{1} + \cdots + (-1)^{r-1}z^{r-2}\binom{r-2}{r-3} + (-1)^r z^{r-1} \\ &= z\left[1 + (-z)\binom{r-2}{1} + \cdots + (-z)^{r-3}\binom{r-2}{r-3} + (-z)^{r-2}\right] \\ &= z(1-z)^{r-2} = (1-z)^{r-2} - (1-z)^{r-1}.\end{aligned}$$

Thus

$$p(z) = -\frac{(1-z)^{r-1}}{r-1} + \frac{(1-z)^r}{r} + c,$$

where c is a constant. From $p(0) = 0$, it follows that $c = \frac{1}{r(r-1)}$ and $p(1) = c = \frac{1}{r(r-1)}$. □

Theorem 9.2 *Let $f_1(z) = f(\frac{1-z}{z+1})$ with $-1 < z \le 1$. If $f_1(z)$ satisfies the following conditions:*

1. *$f_1(z) \in C^r(-1, 1]$ with $r > 2$;*
2. *$\lim_{z\to -1^+} f_1(z) = \lim_{z\to -1^+} f_1'(z) = 0$;*
3. *$\lim_{z\to -1^+} f_1^{(k)}(z)$ exists for each $k = 2, 3, \ldots, r$,*

then the CCR quadrature (9.11) has the accuracy of order $O(n^{-r+2})$.

Proof: In the light of Theorem 9.1, we only require to prove that $f_\alpha(z)/(z+1)^2$, the integrand in (9.11), is in the set $C^{r-2}[-1, 1]$.

Assign $f_1(-1) = 0$, $f_1'(-1) = 0$, and $f_1^{(k)}(-1) = \lim_{z\to -1^+} f_1^{(k)}(z)$, $k = 2, \ldots, r$, then $f_1(z) \in C^r[-1, 1]$. Let

$$f_\alpha(z) = f\left(\frac{\alpha(1-z)}{z+1}\right), \quad -1 < z \le 1.$$

It can be checked that

$$f_\alpha(z) = f_1(u(z)) \quad \text{with} \quad u(z) = \frac{z+1-\alpha(1-z)}{\alpha(1-z)+(z+1)}.$$

Since $\alpha > 0$, the denominator of $u(z)$ is greater than 0 for $z \in [-1, 1]$. More precisely,

$$\min(2, 2\alpha) \le \alpha(1-z) + (z+1) \le \max(2, 2\alpha), \quad z \in [-1, 1].$$

Therefore, $u(z) \in C^\infty[-1, 1]$. It follows that $\lim_{z\to -1^+} f_\alpha^{(k)}(z)$ exists for $k \in \{0, 1, \ldots, r\}$. Assign $f_\alpha^{(k)}(-1) = \lim_{z\to -1^+} f_\alpha^{(k)}(z)$ for $k \in \{0, 1, \ldots, r\}$, then $f_\alpha(z) \in C^r[-1, 1]$. Moreover,

$$\lim_{z\to -1^+} f_\alpha(z) = \lim_{z\to -1^+} f_1(u(z)) = f_1(-1) = 0,$$

and

$$\lim_{z\to -1^+} f_\alpha'(z) = \lim_{z\to -1^+} u'(z) f_1'(u(z)) = u'(-1) f_1'(-1) = 0.$$

Since $f_\alpha(z) \in C^r[-1, 1]$ and $f_\alpha(-1) = f_\alpha'(-1) = 0$, by the Taylor expansion with remainder, we get

$$\begin{aligned}\frac{f_\alpha(z)}{(z+1)^2} &= \frac{f_\alpha''(-1)}{2!} + \frac{f_\alpha^{(3)}(-1)}{3!}(z+1) + \cdots + \frac{f_\alpha^{(r-1)}(-1)}{(r-1)!}(z+1)^{r-3} \\ &+ \frac{1}{(r-1)!(z+1)^2} \int_{-1}^{z} f_\alpha^{(r)}(t)(z-t)^{r-1} dt;\end{aligned}$$

see [5, p. 47] for instance. The rest is to prove that

$$h(z) \equiv \frac{1}{(z+1)^2}\int_{-1}^{z} f_\alpha^{(r)}(t)(z-t)^{r-1}dt \in C^{r-2}(-1,1],$$

and $\lim_{z\to -1^+} h^{(k)}(z)$ exists for each $k \in \{0,1,\ldots,r-2\}$. Since $\int_{-1}^{z} f_\alpha^{(r)}(t)(z-t)^{r-1}dt \in C^r(-1,1]$ and $1/(z+1)^2 \in C^\infty(-1,1]$, for $k \le r$ and $z \in (-1,1]$, we have

$$\begin{aligned}
h^{(k)}(z) &= \sum_{i=0}^{k}\binom{k}{i}\left[\frac{1}{(z+1)^2}\right]^{(i)}\left[\int_{-1}^{z} f_\alpha^{(r)}(t)(z-t)^{r-1}dt\right]^{(k-i)} \\
&= \sum_{i=0}^{k}\binom{k}{i}\frac{(-1)^i(i+1)!}{(z+1)^{i+2}}\cdot\prod_{l=1}^{k-i}(r-l)\cdot\int_{-1}^{z} f_\alpha^{(r)}(t)(z-t)^{r-1-k+i}dt \\
&= \sum_{i=0}^{k}(-1)^i\binom{k}{i}\frac{(i+1)!(r-1)!}{(r-1-k+i)!}\cdot\frac{\int_{-1}^{z} f_\alpha^{(r)}(t)(z-t)^{r-1-k+i}dt}{(z+1)^{i+2}}. \qquad (9.18)
\end{aligned}$$

Therefore $h(z) \in C^r(-1,1]$. To prove that $\lim_{z\to -1^+} h^{(k)}(z)$ exists for each $k \in \{0,1,\ldots,r-2\}$, we first note that

$$\lim_{z\to -1^+}\frac{\int_{-1}^{z} f_\alpha^{(r)}(t)(z-t)^l dt}{(z+1)^i} = 0, \quad i \le l \le r-1. \qquad (9.19)$$

From (9.19) and (9.18) we see that if $k \le r-3$, i.e., $r-k+i-1 \ge 3+i-1 = i+2$, then

$$\lim_{z\to -1^+} h^{(k)}(z) = 0.$$

Finally, note that for $i = 1,2,\ldots,r$,

$$\lim_{z\to -1^+}\frac{\int_{-1}^{z} f_\alpha^{(r)}(t)(z-t)^{i-1}dt}{(z+1)^i} = \lim_{z\to -1^+}\left[\frac{f_\alpha^{(r)}(-1)}{i} + \frac{\int_{-1}^{z} f_\alpha^{(r)}(t)(z-t)^i dt}{i(z+1)^i}\right] = \frac{f_\alpha^{(r)}(-1)}{i}.$$

It follows from (9.18) and (9.17) that

$$\lim_{z\to -1^+} h^{(r-2)}(z) = \sum_{i=0}^{r-2}(-1)^i\binom{r-2}{i}(r-1)!\cdot\frac{f_\alpha^{(r)}(-1)}{i+2} = \frac{(r-2)!f_\alpha^{(r)}(-1)}{r}.$$

The proof of the theorem is completed. □

Consider the function $f(s) = 1/(1+s^2)$. We have $f_1(z) = \frac{(z+1)^2}{(z+1)^2+(1-z)^2} \in C^\infty[-1,1]$. Moreover, $f_1(-1) = 0$ and $f_1'(-1) = \left.\frac{1-z^2}{(z^2+1)^2}\right|_{z=-1} = 0$. Therefore, the CCR quadrature for $\int_0^\infty f(s)ds$ is of high accurate. Our tests show that when $\alpha = 1$, the error in $Q_{40}^{(CCR)}(f)$ is 2.2204e–16.

9.3 Application of CCR to Wiener-Hopf equations

Applying the CCR quadrature (9.11) to the reformulated half-line integral equation (9.2), we get the discretization linear system

$$d_i y_i + \sum_{j=1}^{n} K(t_i - t_j)\frac{2\alpha w_j}{(z_j+1)^2} y_j = g(t_i), \quad i = 1, 2, \ldots, n, \tag{9.20}$$

where $t_i = \frac{\alpha(1-z_i)}{z_i+1}$, y_i are approximations of $y(t_i)$, and

$$d_i = 1 + \int_0^\infty K(t_i - s)ds - \sum_{j=1}^{n} K(t_i - t_j)\frac{2\alpha w_j}{(z_j+1)^2}.$$

Let

$$\mathbf{y}_n = (y_1, y_2, \ldots, y_n)^T, \quad \mathbf{g} = (g(t_1), g(t_2), \ldots, g(t_n))^T, \quad \widehat{K} = [K(t_i - t_j)]_{i,j=1}^n,$$

and

$$Z = \operatorname{diag}\big((z_1+1)^{-2}, \ldots, (z_n+1)^{-2}\big), \quad W = \operatorname{diag}(w_1, \ldots, w_n), \quad D = \operatorname{diag}(d_1, \ldots, d_n).$$

The linear system (9.20) can be rewritten as

$$(D + 2\alpha\widehat{K}ZW)\mathbf{y}_n = \mathbf{g}. \tag{9.21}$$

To investigate the singularities in the kernel function, we introduce rational variable transformations that correspond to the CCR quadrature. Replacing the variables t and s in (9.1) by $\frac{\alpha(1-\tau)}{\tau+1}$ and $\frac{\alpha(1-z)}{z+1}$ respectively, we get the following integral equation

$$Y(\tau) + 2\alpha \int_{-1}^{1} \frac{H(\tau, z)}{(z+1)^2} Y(z)\mathrm{d}z = G(\tau), \qquad -1 < \tau \le 1, \tag{9.22}$$

where

$$H(\tau, z) = K\left(\frac{\alpha(1-\tau)}{\tau+1} - \frac{\alpha(1-z)}{z+1}\right), \quad Y(\tau) = y\left(\frac{\alpha(1-\tau)}{\tau+1}\right), \quad G(\tau) = g\left(\frac{\alpha(1-\tau)}{\tau+1}\right),$$

with $-1 < \tau, z \le 1$.

We note that the function $H(\tau, z)$ has singularities along $z = \tau$ as τ tends to -1. Moreover, as τ becomes closer to -1, the singularity becomes stronger. In fact, from

$$H(\tau, \tau+\delta) - H(\tau, \tau) = K\left(\frac{2\alpha\delta}{(\tau+1)(\tau+\delta+1)}\right) - K(0),$$

we see that when $\tau \approx -1$, a small δ results in a large $\frac{2\alpha\delta}{(\tau+1)(\tau+\delta+1)}$, and it follows that $H(\tau, \tau+\delta) - H(\tau, \tau)$ can be large. The function $\frac{2\alpha H(\tau,z)}{(z+1)^2}$, the kernel of (9.22), has a stronger singularity as τ tends to -1 because of the Jacobian factor $1/(z+1)^2$. Therefore,

$$\frac{H(\tau, \tau+\delta)}{(\tau+\delta+1)^2} - \frac{H(\tau, \tau)}{(\tau+1)^2} = \frac{K\left(\frac{2\alpha\delta}{(\tau+1)(\tau+\delta+1)}\right)}{(\tau+\delta+1)^2} - \frac{K(0)}{(\tau+1)^2}$$

becomes very large for a very small δ as τ tends to -1.

As we did for (9.1), we reformulate equation (9.22) as

$$\left[1+2\alpha\int_{-1}^{1}\frac{H(\tau,z)}{(z+1)^2}\mathrm{d}z\right]Y(\tau)+2\alpha\int_{-1}^{1}\frac{H(\tau,z)}{(z+1)^2}\left[Y(z)-Y(\tau)\right]\mathrm{d}z=G(\tau). \tag{9.23}$$

Applying the Clenshaw-Curtis quadrature (9.9) to $\int_{-1}^{1}\frac{H(\tau,z)}{(z+1)^2}\big[Y(z)-Y(\tau)\big]\mathrm{d}z$, we also obtain (9.21).

Since the factor $1/(z+1)^2$ in the kernel function of (9.22) is independent of τ, we introduce a new function $X(z) = Y(z)/(z+1)^2$ and then subtract the singularities. In other words, we reformulate (9.22) as

$$\left[(\tau+1)^2+2\alpha\int_{-1}^{1}H(\tau,z)\mathrm{d}z\right]X(\tau)+2\alpha\int_{-1}^{1}H(\tau,z)[X(z)-X(\tau)]\mathrm{d}z=G(\tau) \tag{9.24}$$

and then apply the Clenshaw-Curtis quadrature to $\int_{-1}^{1}H(\tau,z)[X(z)-X(\tau)]\mathrm{d}z$. Thus, we obtain another discretization equation

$$\hat{d}_i x_i + 2\alpha\sum_{j=1}^{n}H(\tau,z_j)w_j x_j = G(z_i), \quad i=1,2,\ldots,n,$$

where

$$\hat{d}_i = (z_i+1)^2+2\alpha\int_{-1}^{1}H(z_i,z)dz-2\alpha\sum_{j=1}^{n}w_jH(z_i,z_j).$$

Let $\mathbf{x}_n = (x_1, x_2, \ldots, x_n)^T$ and $\widehat{D} = \mathrm{diag}(\hat{d}_1, \ldots, \hat{d}_n)$. We get the matrix form of the above discretization equation

$$(\widehat{D}+2\alpha\widehat{K}W)\mathbf{x}_n = \mathbf{g}. \tag{9.25}$$

After getting $\mathbf{x}_n$, we can obtain another approximate solution $\tilde{\mathbf{y}}_n = (\tilde{y}_1, \tilde{y}_2, \ldots, \tilde{y}_n)^T$ by using

$$\tilde{y}_i = x_i(z_i+1)^2, \quad i=1,2,\ldots,n. \tag{9.26}$$

We make the following remarks to end this section.

1. In our numerical tests in section 9.4, we evaluate $2\alpha\int_{-1}^{1}H(z_i,z)dz$ as follows. Let $t_i = \frac{\alpha(1-z_i)}{z_i+1}$. Then

$$\begin{aligned}
&2\alpha\int_{-1}^{1}H(z_i,z)dz\\
&=\int_{0}^{\infty}K(t_i-s)\left(\frac{2\alpha}{s+\alpha}\right)^2ds\\
&=\int_{0}^{t_i}K(t_i-s)\left(\frac{2\alpha}{s+\alpha}\right)^2ds+\int_{0}^{\infty}K(-s)\left(\frac{2\alpha}{s+t_i+\alpha}\right)^2ds.
\end{aligned}$$

For the first term, we replace s by $\frac{t_i(z+1)}{2}$ and then use the Clenshaw-Curtis quadrature

(9.9). For the second term, we use the CCR quadrature (9.11). Thus

$$\begin{aligned}
&2\alpha \int_{-1}^{1} H(z_i, z) dz \\
\approx\ & \frac{t_i}{2} \sum_{j=1}^{n} w_j K\Big(\frac{t_i(1-z_j)}{2}\Big)\Big[\frac{2\alpha}{t_i(z_j+1)/2+\alpha}\Big]^2 \\
&+ \sum_{j=1}^{n} \frac{(2\alpha)^3 w_j}{[2\alpha + t_i(z_j+1)]^2} K\Big(\frac{-\alpha(1-z_j)}{z_j+1}\Big) \\
=\ & \sum_{j=1}^{n} \frac{8\alpha^3 w_j}{[2\alpha + t_i(z_j+1)]^2}\Big[t_i K\Big(\frac{t_i(1-z_j)}{2}\Big) + K\Big(\frac{-\alpha(1-z_j)}{z_j+1}\Big)\Big]. \qquad (9.27)
\end{aligned}$$

As to the integral $\int_0^\infty K(t_i - s)\, ds$, we can exactly evaluate it in many applications; see, for instance, all the examples in section 9.4. In case the integral can not be computed exactly, we can obtain a highly accurate approximation in a way similar to (9.27).

2. When the Clenshaw-Curtis quadrature is applied to the integral equations (9.23) and (9.24), the accuracy of the numerical solutions depends on the smoothness of the integrands $H(\tau, z)[Y(z) - Y(\tau)]/(z+1)^2$ and $H(\tau, z)[X(z) - X(\tau)]$, respectively. Since the singularities in $H(\tau, z)$ are weaker than those in $H(\tau, z)/(z+1)^2$, we expect that the numerical solution $\tilde{\mathbf{y}}_n$ obtained from (9.26) is more accurate than $\mathbf{y}_n$ obtained from (9.21). Numerical experiments in section 9.4 verify our prediction.

9.4 Numerical experiments

In this section, we apply the CCR quadrature to Wiener-Hopf equations with the following types of kernels:

- Type 1: Semi-smooth kernels, i.e., $K(s)$ or its derivative $K^{(j)}(s)$ has a singularity at $s = 0$ and is smooth for $s \neq 0$.
- Type 2: Band-limited kernels, i.e., $K(t-s) = P(|t-s|)e^{-|t-s|}$, where P is a polynomial.
- Type 3: Smooth and fast decaying kernels.
- Type 4: Smooth but slowly decaying kernels.

In the following experiments, we set $\alpha = 10$ in (9.23) and (9.24). Since the sizes of the linear systems (9.21) and (9.25) are not large, we use Gaussian elimination method to solve them. We note that Kang et al. have proposed a Nyström-Clenshaw-Curtis (NCC) quadrature for finite-section Wiener-Hopf equations [10, 11]. The NCC quadrature is a highly accurate quadrature which was derived by using a singularity removal technique and the Clenshaw-Curtis quadrature. In the following, we make a brief comparison of the performance of the CCR quadrature with that of the NCC quadrature.

In the following tables, $\mathbf{y}_n$ and $\tilde{\mathbf{y}}_n$ denote the solutions obtained from (9.21) and (9.26) respectively. The symbol $\mathbf{u}_{k,n}^R$ denotes the numerical solution obtained by using the following composite NCC quadrature: partition the truncation interval $[0, R]$ into k subintervals of

Table 9.1: The number of quadrature points and errors in the numerical solutions for Example 1.

n	$\|\mathbf{y}_n - \mathbf{y}\|$	$\|\tilde{\mathbf{y}}_n - \mathbf{y}\|$	$\|\mathbf{u}_{1,n}^{40} - \mathbf{y}\|$	$\|\mathbf{u}_{2,n}^{40} - \mathbf{y}\|$	$\|\mathbf{u}_{2,n}^{80} - \mathbf{y}\|$	$\|\mathbf{u}_{4,n}^{80} - \mathbf{y}\|$
40	3.747e–06	2.892e–06	3.318e–07	9.562e–06	1.337e–04	1.014e–03
80	2.363e–07	1.828e–07	1.145e–11	1.588e–09	3.318e–07	9.562e–06
160	1.472e–08	1.140e–08	2.749e–14	7.740e–15	1.449e–11	1.588e–09
320	9.201e–10	7.118e–10	–	–	2.749e–14	7.742e–15
640	5.751e–11	4.448e–11	–	–	–	–

the same length and use the (n/k)-point NCC quadrature in each subinterval. The symbol $\mathbf{y}$ denotes the n-vector consisting of the values of the true solution at the quadrature points, and $\|\cdot\| = \|\cdot\|_\infty$. When solving (9.21), we compute the exact values for $2\alpha \int_{-1}^{1} \frac{H(\tau,z)}{(z+1)^2} dz$, and when solving (9.25), we compute $2\alpha \int_{-1}^{1} \frac{H(\tau,z)}{(z+1)^2} dz$ by using (9.27). In the following tables, the symbol "–" denotes the case that increasing the number of quadrature points n does not improve the accuracy in the numerical solution.

Example 1 (semi-smooth kernel) [10, Chapter 3, Example 1]:

$$y(t) + \int_0^\infty \frac{1}{1+e^{|t-s|}} y(s)ds = [2 - 2\ln 2 + \ln(1+e^t)]e^{-t}, \quad 0 \le t < \infty.$$

The true solution of the above equation is $y(t) = e^{-t}$.

The errors in the numerical solutions are listed in Table 9.1. For the NCC method, as the true solution decays exponentially, there is no need to choose a very large R. We observe that:

1. With the same number of quadrature points, $\tilde{\mathbf{y}}_n$ is more accurate than $\mathbf{y}_n$, i.e., the numerical solution obtained from (9.26) is better than the one obtained from (9.21).

2. The accuracy of $\mathbf{y}_n$ and $\tilde{\mathbf{y}}_n$ is about $O(n^{-4})$.

3. The truncation $R = 40$ is large enough for this example and the NCC method is more efficient than the CCR method. The number of subintervals does not need to be too large.

Example 2 (band-limited kernel) [10, Chapter 3, Example 2]:

$$y(t) + \int_0^\infty K(t-s)y(s)ds = (2 + t + t^2/2 + t^3/3)e^{-t}, \quad 0 \le t < \infty,$$

where $K(t) = \left(1 + |t| + |t|^2\right)e^{-|t|}$. The true solution of the above equation is $y(t) = e^{-t}$.

It is easy to judge that the kernel of Example 2 is also of semi-smooth type. From Table 9.2 we observe that with the CCR method, the accuracy of the numerical solution is about $O(n^{-6})$. Again, $\tilde{\mathbf{y}}_n$ is better than $\mathbf{y}_n$. Similar to Example 1, we do not need a large R. We notice that, due to the factor $e^{-|t-s|}$ in the kernel (in the NCC method, it is required that both e^{s-t} and e^{t-s} are bounded), we have to partition the interval $[0, R]$ into subintervals of small length. Otherwise, there exist entries in the coefficient matrix which are very large.

Table 9.2: The number of quadrature points and errors in the numerical solutions for Example 2.

n	$\|\mathbf{y}_n - \mathbf{y}\|$	$\|\tilde{\mathbf{y}}_n - \mathbf{y}\|$	$\|\mathbf{u}_{4,n}^{40} - \mathbf{y}\|$	$\|\mathbf{u}_{8,n}^{40} - \mathbf{y}\|$	$\|\mathbf{u}_{4,n}^{80} - \mathbf{y}\|$	$\|\mathbf{u}_{8,n}^{80} - \mathbf{y}\|$
40	9.978e–07	7.893e–07	0.156	1.093	0.881	0.778
80	1.589e–08	1.283e–08	5.803e–06	2.666e–03	0.971	0.156
160	2.507e–10	1.997e–10	1.552e–06	1.167e–10	1.800e–05	5.803e–06
320	3.913e–12	3.121e–12	–	1.725e–13	–	1.552e–10
640	6.156e–14	4.971e–14	–	–	–	–

Table 9.3: The number of quadrature points and errors in the numerical solutions for Example 3.

n	$\|\mathbf{y}_n - \mathbf{y}\|$	$\|\tilde{\mathbf{y}}_n - \mathbf{y}\|$	$\|\mathbf{u}_{1,n}^{40} - \mathbf{y}\|$	$\|\mathbf{u}_{2,n}^{40} - \mathbf{y}\|$	$\|\mathbf{u}_{1,n}^{80} - \mathbf{y}\|$	$\|\mathbf{u}_{2,n}^{80} - \mathbf{y}\|$
40	1.284e–04	1.016e–04	5.236e–04	2.492e–03	8.429e–03	1.268
80	3.447e–06	2.921e–06	7.390e–07	3.098e–06	5.684e–05	5.236e–04
160	2.985e–08	2.609e–08	–	7.395e–07	1.561e–08	3.975e–07
320	5.689–11	5.147e–11	–	–	1.198e–12	1.301e–12
640	1.491e–14	1.375e–14	–	–	–	1.198e–12

From Table 9.2, we see that for most cases, the CCR method performs better than the NCC method.

Example 3 (smooth and fast decaying kernel) [3, Example 1]:

$$y(t) - \int_0^\infty \frac{\sqrt{3}}{\pi(e^{t-s} + e^{s-t})} y(s)ds = g(t), \quad 0 \le t < \infty,$$

where

$$g(t) = e^{-t/3}\Big(\frac{1}{4} + \frac{\sqrt{3}}{2\pi}\ln\frac{u+1}{\sqrt{u^2-u+1}} + \frac{3}{2\pi}\arctan\frac{2u-1}{\sqrt{3}}\Big)$$

with $u = e^{-2t/3}$. The true solution is $y(t) = e^{-t/3}$.

We observe from Table 9.3 that as n increases, the errors in $\mathbf{y}_n$ and $\tilde{\mathbf{y}}_n$ decay rather rapidly. In fact, the convergence is superalgebraic. On the other hand, since the exact solution does not decay as fast as the solutions in examples 1 and 2, a larger R should be chosen to avoid large truncation error. Since the kernel function is smooth, we set k to be small integers. From the numerical results, we see that $R = 80$ is a suitable choice. Furthermore, for most n's, $\mathbf{y}_n$ and $\tilde{\mathbf{y}}_n$ are more accurate than $\mathbf{u}_{k,n}^R$.

Example 4 (smooth but slowly decaying kernel):

$$y(t) + \int_0^\infty \frac{1}{1+(t-s)^2} y(s)ds = g(t), \quad 0 \le t < \infty,$$

Table 9.4: The number of quadrature points and errors in the numerical solutions for Example 4.

n	$\|\mathbf{y}_n - \mathbf{y}\|$	$\|\tilde{\mathbf{y}}_n - \mathbf{y}\|$	$\|\mathbf{u}_{1,n}^{160} - \mathbf{y}\|$	$\|\mathbf{u}_{2,n}^{160} - \mathbf{y}\|$	$\|\mathbf{u}_{1,n}^{640} - \mathbf{y}\|$	$\|\mathbf{u}_{2,n}^{640} - \mathbf{y}\|$
40	6.024e–06	2.639e–06	4.405e–03	4.619e–02	2.751e–02	0.163
80	4.998e–07	1.219e–07	2.568e–04	1.054e–03	4.515e–03	4.687e–02
160	5.655e–08	8.552e–09	3.899e–05	4.196e–05	2.803e–04	1.144e–03
320	7.373e–09	7.323e–10	–	3.905e–05	1.542e–05	6.256e–05
640	1.039e–09	6.682e–11	–	–	2.599e–06	2.534e–06

where

$$g(t) = \frac{1}{1+t^2} + \frac{1}{4+t^2}\big[\pi + \arctan(t)\big] + \frac{\ln(1+t^2)}{t(4+t^2)}.$$

The true solution of the equation is $y(t) = 1/(1+t^2)$.

We observe from Table 9.4 that $\tilde{\mathbf{y}}_n$ is much more accurate than $\mathbf{y}_n$, and the errors in $\mathbf{y}_n$ and $\tilde{\mathbf{y}}_n$ are $O(n^{-2.7})$ and $O(n^{-3.5})$ respectively. With 640 quadrature points, $\tilde{\mathbf{y}}_n$ achieves the accuracy of 6.6818e–11, while the accuracy of $\mathbf{y}_n$ is only 1.0388e–09. Since the exact solution decays quite slowly, to avoid large truncation error, a very large R should be chosen and the performance of the NCC method is not good. We can see from Table 9.4 that for all n, $\mathbf{y}_n$ and $\tilde{\mathbf{y}}_n$ are much more accurate than $\mathbf{u}_{k,n}^R$.

To end this section, we compare numerically the singularities of the integrands in (9.22), (9.23), and (9.24), i.e.,

$$h_{1,\tau}(z) = \frac{H(\tau,z)Y(z)}{(z+1)^2}, \quad h_{2,\tau}(z) = \frac{H(\tau,z)[Y(z)-y(\tau)]}{(z+1)^2}, \quad h_{3,\tau}(z) = H(\tau,z)[X(z)-X(\tau)],$$

respectively. The curves of the integrands for three τ's: $\cos(39\pi/640) \approx 0.9257$ (close to 1), $\cos(319\pi/640) \approx 0.0049$ (close to 0), and $\cos(559\pi/640) \approx -0.9220$ (close to -1), are shown in Figures 9.1–9.3, respectively. The integrands are from Example 4, whose kernel function $K(t-s)$ is smooth. We can see from Figure 9.1 that for $\tau = \cos(39\pi/640)$ (close to 1), all integrands, $h_{i,\tau}(z)$ $(i = 1,2,3)$, are very smooth. From Figure 9.2, we observe that for $\tau = \cos(319\pi/640)$ (close to 0), there is a considerable singularity in $h_{1,\tau}(z) = H(\tau,z)Y(z)/(z+1)^2$, while the other two integrands look quite smooth. Figure 9.3 shows the case where $\tau = \cos(559\pi/640)$ (close to -1). In this case, all the integrands have singularities: $h_{1,\tau}(z)$ has a stronger singularity at $z = \tau$, the singularity in $h_{2,\tau}(z)$ is weaker, and the one in $h_{3,\tau}(z)$ is the weakest.

From the figures, we also observe that when τ is close to -1, the value of $h_{1,\tau}(z)$ is much larger than that of $h_{2,\tau}(z)$ and $h_{3,\tau}(z)$. In summary, comparing to $h_{2,\tau}(z)$ and $h_{3,\tau}(z)$, $h_{1,\tau}(z)$ has larger function value and much stronger singularity. Therefore, the errors in $Q_n(h_{1,\tau})$ are much larger than those in $Q_n(h_{2,\tau})$ and $Q_n(h_{3,\tau})$.

9.5 Concluding remarks

In this chapter, we proposed a CCR quadrature rule which is based on the Clenshaw-Curtis quadrature and a rational variable transformation. Then we apply it to solve Wiener-Hopf

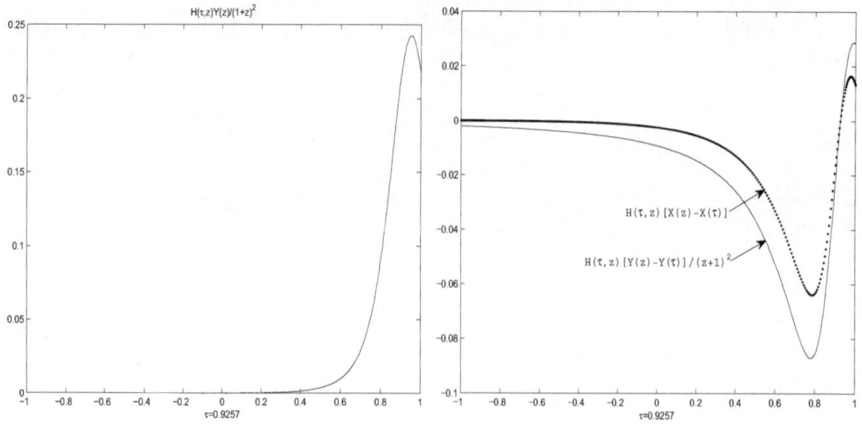

Figure 9.1: Curves of the integrands in (9.22), (9.23), and (9.24): $\tau = \cos\left(\frac{39\pi}{640}\right) \approx 0.9257$.

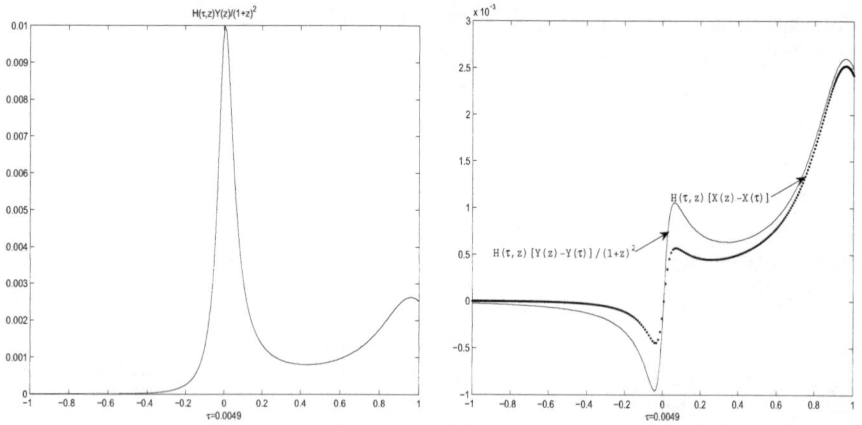

Figure 9.2: Curves of the integrands in (9.22), (9.23), and (9.24): $\tau = \cos\left(\frac{319\pi}{640}\right) \approx 0.0049$.

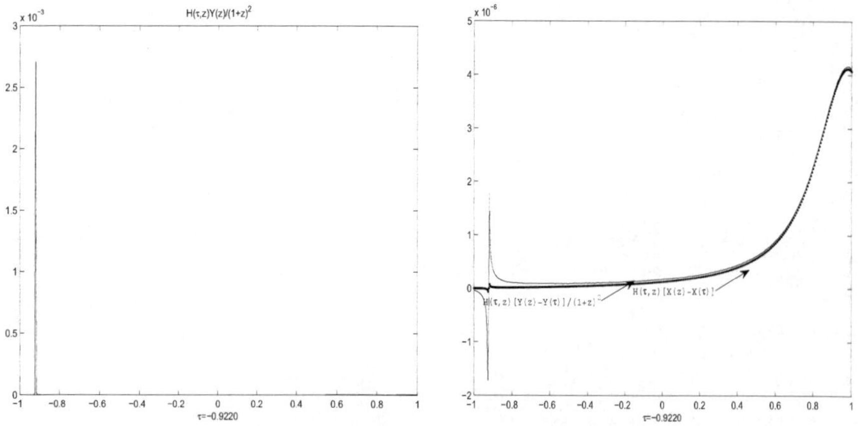

Figure 9.3: Curves of the integrands in (9.22), (9.23), and (9.24): $\tau = \cos\left(\frac{559\pi}{640}\right) \approx -0.9220$.

equations of the second kind on the half-line. Numerical examples with several typical kinds of kernels are given. Our results indicate that the CCR method is well suited not only for integrands with exponential decay but also for those that decay quadratically.

References for Chapter 9

[1] P. M. Anselone and C. T. H. Baker, *Error bounds for integral equations on the half line*, J. Integral Equations Appl., 1 (1988), pp. 321–324.

[2] P. M. Anselone and I. H. Sloan, *Integral equations on the half line* (II): *the Wiener-Hopf case*, J. Integral Equations Appl., 1 (1988), pp. 203–225.

[3] G. A. Chandler and I. G. Graham, *The convergence of Nyström methods for Wiener-Hopf equations*, Numer. Math., 2 (1988), pp. 345–364.

[4] C. W. Clenshaw and A. R. Curtis, *A method for numerical integration on an automatical computer*, Numer. Math., 197 (1960), pp. 197–205.

[5] L. M. Delves and J. L. Mohamed, *Computational Methods for Integral Equations*, Cambridge University Press, Cambridge 1985.

[6] I. C. Gohberg and I. A. Fel'dman, *Convolution Equations and Projection Methods for Their Solution*, Transl. Maths. Monos. 41, Amer. Maths. Soc., Providence, RI, 1974.

[7] I. Gohberg, M. Hanke, and I. Koltracht, *Fast preconditioned conjugate gradient algorithms for Wiener-Hopf integral equations*, SIAM J. Numer. Anal., 31 (1994), pp. 429–443.

[8] I. G. Graham and W. R. Mendes, *Nyström-product integration for Wiener-Hopf equations with applications to radiative transfer*, IMA J. Numer. Anal., 9 (1989), pp. 261–284.

[9] C. D. Green, *Integral Equation Methods*, Thomas & Sons Ltd, Nelson, London, 1969.

[10] S. Y. Kang, *Numerical Solution of Integral Equation with Nonsmooth Kernels and Applications,* PhD thesis, Department of Mathematics, University of Connecticut, Storrs, CT, 2000.

[11] S. Y. Kang, I. Koltracht, and G. Rawitscher, *Nyström-Clenshaw-Curtis quadrature for integral equations with discontinuous kernels*, Math. Comp., 72 (2003), pp. 729–756.

[12] F. R. Lin, M. K. Ng, and R. H. Chan, *Preconditioner for Wiener-Hopf equations with high order quadrature rules*, SIAM J. Numer. Anal., 34 (1997), pp. 1418–1431.

[13] G. Mastroianni and G. Monegato, *Nyström interpolants based on zeros of Laguerre polynomials for some Wiener-Hopf equations*, IMA J. Numer. Anal., 17 (1997), pp. 621–642.

[14] I. H. Sloan and A. Spence, *Projection methods for integral equations on the half line*, IMA J. Numer. Anal., 6 (1986), pp. 153–172.

[15] I. H. Sloan and A. Spence, *Integral equations on the half line: a modified finite section approximation*, Math. Comp., 47 (1986), pp. 589–595.

[16] L. N. Trefethen, *Is Gauss quadrature better than Clenshaw-Curtis?*, SIAM Review, 50 (2008), pp. 67–87.

Chapter 10

On the Solution of Singular Systems by Krylov Subspace Methods

Man-chung Yeung [1]

Abstract

Krylov subspace methods are popular iterative methods to solve large sparse linear systems in the real-world computations due to their cheap memory requirement and computational cost. In this chapter, we discuss the solution of singular systems. We show that the consistency of a singular linear system is not a sufficient condition for a Krylov subspace method to successfully find a solution to the system. The choice of an initial guess is a crucial step. If the initial guess is properly chosen, a Krylov subspace method almost surely converges to find a solution from a probabilistic point of view. Otherwise, a Krylov subspace method surely diverges.

10.1 Introduction

Krylov subspace methods are iterative methods for the solution of linear systems

$$A\mathbf{x} = \mathbf{b}, \tag{10.1}$$

where $A \in \mathbb{R}^{N\times N}$ (singular or nonsingular) and $\mathbf{b} \in \mathbb{R}^N$. At the kth iteration, a Krylov subspace method searches for an approximate solution $\mathbf{x}_k$ from the affine space

$$\mathbf{x}_0 + \text{span}\left\{\mathbf{r}_0, A\mathbf{r}_0, \ldots, A^{k-1}\mathbf{r}_0\right\} \equiv \mathbf{x}_0 + \mathcal{K}_k(A, \mathbf{r}_0),$$

[1]Department of Mathematics, University of Wyoming, 1000 East University Avenue, Laramie, WY 82071, USA (email: myeung@uwyo.edu).
This research was partially supported by Flittie Sabbatical Augmentation Award, University of Wyoming.

where $\mathbf{x}_0 \in \mathbb{R}^N$ is an initial guess to the solution of (10.1), $\mathbf{r}_0 = \mathbf{b} - A\mathbf{x}_0$ is the residual of $\mathbf{x}_0$, and $\mathcal{K}_k(A, \mathbf{r}_0) = \text{span}\{\mathbf{r}_0, A\mathbf{r}_0, \ldots, A^{k-1}\mathbf{r}_0\}$ is called a Krylov subspace.

Suppose that $\mathbf{x}_*$ is a solution to (10.1). If A is nonsingular, it is obvious that

$$\mathbf{x}_* \in \mathbf{x}_0 + \text{span}\{\mathbf{r}_0, A\mathbf{r}_0, \ldots, A^{N-1}\mathbf{r}_0\} \tag{10.2}$$

for any $\mathbf{x}_0 \in \mathbb{R}^N$. Therefore, $\mathbf{x}_*$ can be found by a Krylov subspace method within N iterations. Nevertheless, if A is singular, there is no guarantee that (10.2) holds for an arbitrary $\mathbf{x}_0$. For example, the affine space $\mathbf{x}_0 + \mathcal{K}_N(A, \mathbf{r}_0)$ contains no solution of $A\mathbf{x} = \mathbf{b}$ if

$$A = \begin{bmatrix} 0 & 1 \\ 0 & 0 \end{bmatrix} \quad \text{and} \quad \mathbf{b} = \begin{bmatrix} 1 \\ 0 \end{bmatrix}$$

with $\mathbf{x}_0 = (1, 0)^T$ being selected. Note that this linear system is consistent. Thus the selection of $\mathbf{x}_0$ is crucial for a Krylov subspace method to solve a singular system (10.1). If $\mathbf{x}_0$ is chosen such that the affine space $\mathbf{x}_0 + \mathcal{K}_N(A, \mathbf{r}_0)$ contains a solution to $A\mathbf{x} = \mathbf{b}$, then the Krylov subspace method converges to find a solution; otherwise, the Krylov subspace method diverges.

Usually, there are some divisions in the algorithm of a Krylov subspace method. If, at some iteration, the denominator of a division becomes zero, then the algorithm will have to stop there and can not continue to search for the solution. Computationally we call this situation *breakdown by zero.*

In this chapter, we show that when an initial guess $\mathbf{x}_0$ is chosen such that (10.2) holds, a Krylov subspace method will almost surely encounter no breakdown and converge to find a solution to (10.1).

10.2 ML(n)BiCG method

ML(n)BiCG is a Krylov subspace method derived by Yeung and Chan in [6]. This method was built upon a Lanczos process with n left starting vectors and a single right starting vector. We choose ML(n)BiCG to study because it contains many other standard methods as its special cases.

Algorithm 10.1 ML(n)BiCG

1. *Choose an initial guess* $\mathbf{x}_0$ *and* n *vectors* $\mathbf{q}_1, \mathbf{q}_2, \ldots, \mathbf{q}_n$.
2. *Compute* $\mathbf{r}_0 = \mathbf{b} - A\mathbf{x}_0$ *and set* $\mathbf{p}_1 = \mathbf{q}_1$, $\mathbf{g}_0 = \mathbf{r}_0$.
3. *For* $k = 1, 2, \ldots,$ *until convergence*:
4. $\quad \alpha_k = \dfrac{\mathbf{p}_k^* \mathbf{r}_{k-1}}{\mathbf{p}_k^* A \mathbf{g}_{k-1}}$;
5. $\quad \mathbf{x}_k = \mathbf{x}_{k-1} + \alpha_k \mathbf{g}_{k-1}$;
6. $\quad \mathbf{r}_k = \mathbf{r}_{k-1} - \alpha_k A \mathbf{g}_{k-1}$;
7. $\quad$ *For* $s = \max(k-n, 0), \ldots, k-1$
8. $\quad\quad \beta_s^{(k)} = -\dfrac{\mathbf{p}_{s+1}^* A \left(\mathbf{r}_k + \sum_{t=\max(k-n,0)}^{s-1} \beta_t^{(k)} \mathbf{g}_t\right)}{\mathbf{p}_{s+1}^* A \mathbf{g}_s}$;
9. $\quad$ *End*
10. $\quad \mathbf{g}_k = \mathbf{r}_k + \sum_{s=\max(k-n,0)}^{k-1} \beta_s^{(k)} \mathbf{g}_s$;
11. $\quad$ *Compute* $\mathbf{p}_{k+1} = (A^*)^j \mathbf{q}_i$, *where* $j \geq 0, 1 \leq i \leq n$, *and* $k + 1 = jn + i$.
12. *End*

In Algorithm 10.1, the vectors $\mathbf{q}_1, \ldots, \mathbf{q}_n$ at step 1 can be any linearly independent vectors.

10.2.1 Relations to some existing methods

ML(n)BiCG contains BiCG and FOM methods as its special cases.

1. *Relation with BiCG* [2]. When $n = 1$, we have $\mathbf{p}_{k+1} = (A^*)^k \mathbf{q}_1$ at step 11 in Algorithm 10.1. By Proposition 10.1 (a) and (d) in section 10.2.2, $\mathbf{x}_k$ and $\mathbf{r}_k$ computed by Algorithm 10.1 satisfy

$$\begin{cases} \mathbf{x}_k \in \mathbf{x}_0 + \text{span}\left\{\mathbf{r}_0, A\mathbf{r}_0, \ldots, A^{k-1}\mathbf{r}_0\right\}, \\ \mathbf{r}_k \perp \text{span}\left\{\mathbf{q}_1, A^*\mathbf{q}_1, \ldots, (A^*)^{k-1}\mathbf{q}_1\right\}, \end{cases} \tag{10.3}$$

for $1 \leq k \leq \nu$. The property (10.3) is what the BiCG approximate solution $\mathbf{x}_k^{BiCG}$ needs to satisfy. Therefore, when $n = 1$, Algorithm 10.1 is mathematically equivalent to the BiCG.

2. *Relation with FOM* [5]. When $n \geq N$, we have $\mathbf{p}_{k+1} = \mathbf{q}_{k+1}$ at step 11 in Algorithm 10.1. If we choose $\mathbf{q}_{k+1} = \mathbf{r}_k$ in Algorithm 10.1 (it is possible since $\mathbf{r}_k$ is computed before $\mathbf{q}_{k+1}$ is used), then by Proposition 10.1 (a) and (d) in section 10.2.2 again, $\mathbf{x}_k$ and $\mathbf{r}_k$ computed by the algorithm satisfy

$$\begin{cases} \mathbf{x}_k \in \mathbf{x}_0 + \text{span}\left\{\mathbf{r}_0, A\mathbf{r}_0, \ldots, A^{k-1}\mathbf{r}_0\right\}, \\ \mathbf{r}_k \perp \text{span}\left\{\mathbf{r}_0, \mathbf{r}_1, \ldots, \mathbf{r}_{k-1}\right\}, \end{cases} \tag{10.4}$$

for $1 \leq k \leq N$. The property (10.4) is what the FOM approximate solution $\mathbf{x}_k^{FOM}$ needs to satisfy. Therefore, when $n \geq N$ and with the choice $\mathbf{q}_{k+1} = \mathbf{r}_k$, Algorithm 10.1 is mathematically equivalent to the FOM.

10.2.2 Properties of ML(n)BiCG

Let ν be the degree of the minimal polynomial $p_{\min}(\lambda, A, \mathbf{r}_0)$ of $\mathbf{r}_0$ with respect to A (i.e., the unique monic polynomial $p(\lambda)$ of minimum degree such that $p(A)\mathbf{r}_0 = 0$). Furthermore, we set

$$S_\nu = [\mathbf{p}_1, \mathbf{p}_2, \ldots, \mathbf{p}_\nu]^* \left[A\mathbf{r}_0, A^2\mathbf{r}_0, \ldots, A^\nu\mathbf{r}_0\right]$$

and

$$W_\nu = [\mathbf{p}_1, \mathbf{p}_2, \ldots, \mathbf{p}_\nu]^* \left[\mathbf{r}_0, A\mathbf{r}_0, \ldots, A^{\nu-1}\mathbf{r}_0\right].$$

Let S_l and W_l denote the $l \times l$ leading principal submatrices of S_ν and W_ν respectively. We now summarize some useful facts about Algorithm 10.1. They can be observed from the construction procedure of the algorithm.

Proposition 10.1 *In exact arithmetic, if*

$$\prod_{l=1}^{\nu} \det(S_l)\det(W_l) \neq 0, \tag{10.5}$$

then Algorithm 10.1 *does not break down by zero division for* $k = 1, 2, \ldots, \nu$, *and* $\mathbf{x}_\nu$ *is the exact solution of* (10.1). *Moreover, the computed quantities satisfy*

(a) $\mathbf{x}_k \in \mathbf{x}_0 + \mathcal{K}_k(A, \mathbf{r}_0)$ *and* $\mathbf{r}_k = \mathbf{b} - A\mathbf{x}_k \in \mathbf{r}_0 + A\mathcal{K}_k(A, \mathbf{r}_0)$ *for* $1 \le k \le \nu$.

(b) $\mathrm{span}\{\mathbf{r}_0, \mathbf{r}_1, \ldots, \mathbf{r}_{k-1}\} = \mathcal{K}_k(A, \mathbf{r}_0)$ *for* $1 \le k \le \nu$.

(c) $\mathrm{span}\{A\mathbf{r}_0, A\mathbf{r}_1, \ldots, A\mathbf{r}_{\nu-1}\} = \mathcal{K}_\nu(A, \mathbf{r}_0)$.

(d) $\mathbf{r}_k \perp \mathrm{span}\{\mathbf{p}_1, \mathbf{p}_2, \ldots, \mathbf{p}_k\}$ *and* $\mathbf{r}_k \not\perp \mathbf{p}_{k+1}$ *for* $0 \le k \le \nu - 1$.

Remark 10.1

(i) *The matrices* S_l *and* W_l *have already appeared in* [3, 4] *where they were called moment matrices. Proposition* 10.1 *can be regarded as a generalization of Theorem* 2 *in* [4] *from* $n = 1$ *to* $n > 1$.

(ii) *ML(n)BiCG also has two types of breakdown caused by the failure of the underlying Lanczos process and the nonexistence of the LU factorizations of the Hessenberg matrix of the recurrence coefficients respectively. Both types of breakdown are reflected in Algorithm* 10.1 *by the denominator* $\mathbf{p}_k^* A\mathbf{g}_{k-1} = 0$. *The condition* $\prod_{l=1}^{\nu} \det(W_l) \neq 0$ *guarantees that the underlying Lanczos process works without breakdown, and the condition* $\prod_{l=1}^{\nu} \det(S_l) \neq 0$ *ensures that the LU factorizations exist.*

(iii) *Notice that* $\det(S_\nu) \neq 0$ *implies* $p_{\min}(0, A, \mathbf{r}_0) \neq 0$ *which, in turn, implies that* (10.1) *is consistent and* (10.2) *holds.*

10.3 Solution of singular systems

Proposition 10.1 indicates that ML(n)BiCG will find a solution to (10.1) if the condition (10.5) holds, no matter (10.1) is singular or not. In this section, we prove that (10.5) holds if and only if (10.2) holds from a probabilistic viewpoint.

Lemma 10.1 *Consider the case that* $n = 1$, $\mathbf{r}_0 \in \mathbb{R}^N$, $\mathbf{r}_0 \neq 0$, *and* $A \in \mathbb{R}^{N\times N}$ *is nonsingular. If* $\mathbf{q}_1 \in \mathbb{R}^N$ *is a random vector with independent and identically distributed elements from* $\mathcal{N}(0, 1)$, *the normal distribution with mean* 0 *and variance* 1, *then*

$$\mathrm{Prob}\Big(\prod_{l=1}^{\nu} \det(W_l)\det(S_l) = 0\Big) = 0.$$

The proof of the lemma requires some properties of Gaussian matrices described in [1]. The matrix A in Lemma 10.1 is assumed to be nonsingular. For a general A, we have the following theorem.

Theorem 10.1 *Consider the case that* $n = 1$, $\mathbf{r}_0 \in \mathbb{R}^N$, $\mathbf{r}_0 \neq 0$, *and* $A \in \mathbb{R}^{N\times N}$. *If* $\mathbf{q}_1 \in \mathbb{R}^N$ *is a random vector with independent and identically distributed elements from* $\mathcal{N}(0, 1)$, *then* $\mathrm{Prob}\Big(\prod_{l=1}^{\nu} \det(W_l)\det(S_l) = 0\Big) = 0$ *if and only if* $p_{\min}(0, A, \mathbf{r}_0) \neq 0$.

Extension of the theorem to the general case is possible, namely, $n \geq 1, A \in \mathbb{R}^{N\times N}$, $\mathbf{r}_0 \in \mathbb{R}^N$, and $[\mathbf{q}_1, \ldots, \mathbf{q}_n]$ is a Gaussian matrix.

Remark 10.2 *The condition $p_{\min}(0, A, \mathbf{r}_0) \neq 0$ holds if and only if $\mathbf{x}_0 + \mathcal{K}_N(A, \mathbf{r}_0)$ contains a solution to* (10.1), *namely,* (10.2) *holds.*

The following corollary then follows from Proposition 10.1 and Theorem 10.1.

Corollary 10.1 *In the case that $n = 1$ and $\mathbf{q}_1 \in \mathbb{R}^N$ is a random vector with independent and identically distributed elements from $\mathcal{N}(0,1)$, Algorithm* 10.1 *almost surely works without breakdown by zero division to find a solution from the affine space $\mathbf{x}_0 + \mathcal{K}_N(A, \mathbf{r}_0)$ provided that $\mathbf{x}_0 \in \mathbb{R}^N$ is chosen such that the affine space contains a solution to* (10.1).

Remark 10.3

(i) *The initial guess $\mathbf{x}_0$ in Corollary* 10.1 *is a user-provided vector. It may not be a random vector in some applications. For example, in the case that a sequence of similar linear systems is solved, the solution from the previous system may be used as the $\mathbf{x}_0$ for the new system.*

(ii) *If we pick $\mathbf{x}_0 \in \mathbb{R}^N$ randomly and set $\mathbf{q}_1 = \mathbf{b} - A\mathbf{x}_0$, then Algorithm* 10.1 *with $n = 1$, or equivalently in mathematics, the standard BiCG (see section* 10.2.1), *almost surely solves* (10.1) *without breakdown by zero division for all, but a certain small class of, nonsingular $A \in \mathbb{R}^{N\times N}$. For details, see* [4].

10.4 Conclusion

When a singular system is solved by Krylov subspace methods, selecting an initial guess $\mathbf{x}_0$ is crucial. If $\mathbf{x}_0$ is selected such that the affine space $\mathbf{x}_0 + \text{span}\{\mathbf{r}_0, A\mathbf{r}_0, \cdots, A^{N-1}\mathbf{r}_0\}$ contains a solution to the system $A\mathbf{x} = \mathbf{b}$, then ML(n)BiCG will almost surely converge (see Corollary 10.1). Otherwise, we have $p_{\min}(0, A, \mathbf{r}_0) = 0$ (see the remark right before Corollary 10.1) which yields $\det(S_\nu) = 0$ and hence $\prod_{l=1}^{\nu} \det(S_l) = 0$. In this case, there is no guarantee that the LU-factorizations in the construction of ML(n)BiCG exist (see the remark right after Proposition 10.1). As a result, it is likely that $\|\mathbf{r}_k\|_2$ blows up to ∞. Although we have picked ML(n)BiCG as a representative to study, a similar conclusion is also true for other Krylov subspace methods.

References for Chapter 10

[1] A. Edelman, *Eigenvalues and condition numbers of random matrices,* SIAM J. Matrix Anal. Appl., 9 (1988), pp. 543–560.

[2] R. Fletcher, *Conjugate gradient methods for indefinite systems,* volume 506 of Lecture Notes Math., pp. 73–89. Springer-Verlag, Berlin-Heidelberg-New York, 1976.

[3] W. D. Joubert, *Generalized Conjugate Gradient and Lanczos Methods for the Solution of Nonsymmetric Systems of Linear Equations*, Ph.D. thesis and Tech. Report CNA-238, Center for Numerical Analysis, University of Texas, Austin, TX, 1990.

[4] W. D. Joubert, *Lanczos methods for the solution of nonsymmetric systems of linear equations*, SIAM J. Matrix Anal. Appl., 13 (1992), pp. 926–943.

[5] Y. Saad and M. H. Schultz, *GMRES: A generalized minimal residual algorithm for solving nonsymmetric linear systems,* SIAM J. Sci. Statist. Comput., 7 (1986), pp. 856–869.

[6] M. Yeung and T. Chan, *ML(k)BiCGSTAB: A BiCGSTAB variant based on multiple Lanczos starting vectors*, SIAM J. Sci. Comput., 21 (1999), pp. 1263–1290.

Chapter 11

A Comparative Study of the LLL Algorithm

Qi-fang Yu [1] **San-zheng Qiao** [2] **Yi-min Wei** [3]

Abstract

When solving the integer least squares problems arise in many engineering applications, the LLL algorithm is widely used as a preconditioner. However, the way how the LLL algorithm works was seldom understood by practitioners, until Luk and Tracy [4] first describe the behavior of the algorithm. In [4], they also derive a new numerical implementation of the LLL algorithm. Luk and Qiao [3] compare this new LLL algorithms and the original one in the cases of overflow and underflow. In this chapter, we look into the difference when they are applied to solve integer least squares problems.

11.1 Introduction

Consider the integer least squares problem:

$$\min_{\mathbf{x}\in\mathbb{Z}^n} \|A\mathbf{x} - \mathbf{b}\|_2^2, \tag{11.1}$$

[1]Institute of Mathematics, School of Mathematical Science, Fudan University, & Shanghai Key Laboratory of Contemporary Applied Mathematics, Shanghai, 200433, P. R. China (email: 072018023@fudan.edu.cn). This author was supported by KLMM0901.

[2]Department of Computing and Software, McMaster University, Hamilton, Ontario, Canada L8S 4K1 (email: qiao@mcmaster.ca). This author was partially supported by Key Laboratory of Nonlinear Science (Fudan University), Ministry of Education.

[3]Institute of Mathematics, School of Mathematical Science, Fudan University, & Shanghai Key Laboratory of Contemporary Applied Mathematics, Shanghai, 200433, P. R. China (emails: ymwei@fudan.edu.cn and yimin.wei@gmail.com). This author was supported by the National Natural Science Foundation of China under grant 10871051, Doctoral Program of the Ministry of Education under grant 20090071110003, Shanghai Municipal Education Committee (Dawn Project), and Shanghai Municipal Science & Technology Committee under grant 09DZ2272900.

where $A \in \mathbb{R}^{m\times n}$ with $m \geq n$ and $\mathbf{b} \in \mathbb{R}^m$. Assuming that A is of full column rank, in a standard method for solving the problem, the matrix A is first reduced to upper triangular by the QR decomposition

$$A = P\begin{bmatrix} R \\ 0 \end{bmatrix},$$

where P is orthogonal and R is upper triangular. Partitioning $P = [P_1, P_2]$ according to R, i.e., P_1 is m-by-n, then we have

$$\|A\mathbf{x} - \mathbf{b}\|_2^2 = \left\| \begin{bmatrix} R \\ 0 \end{bmatrix} \mathbf{x} - [P_1\ P_2]^{\mathrm{T}}\mathbf{b} \right\|_2^2 = \|R\mathbf{x} - \widehat{\mathbf{b}}\|_2^2 + \|P_2^{\mathrm{T}}\mathbf{b}\|_2^2,$$

where $\widehat{\mathbf{b}} = P_1^{\mathrm{T}}\mathbf{b}$. Thus the general integer least squares problem (11.1) is reduced to the triangular integer least squares problem:

$$\min_{\mathbf{x}\in\mathbb{Z}^n} \|R\mathbf{x} - \widehat{\mathbf{b}}\|_2^2. \tag{11.2}$$

The complexity of searching for the integer least squares solution in the entire lattice space is non-polynomial. To reduce the complexity, in the sphere decoding method, the triangular least squares problem (11.2) is constrained to the lattice points in a sphere $\mathcal{S} = \{\mathbf{x} \mid \|R\mathbf{x} - \widehat{\mathbf{b}}\|_2 \leq \rho\}$ for a given radius ρ:

$$\min_{\mathbf{x}\in\mathcal{S}} \|R\mathbf{x} - \widehat{\mathbf{b}}\|_2^2.$$

The LLL algorithm is used as a preconditioner for solving the above problem. It is shown that it can reduce both the radius ρ and the total number of search paths [5].

The LLL algorithm, originated from [2], has many important applications, such as cryptography and GPS in addition to the sphere decoding in wireless communication [1]. It computes the QRZ decomposition:

$$R = Q\widehat{R}Z^{-1}, \tag{11.3}$$

where Q is orthogonal and Z is an integer unimodular matrix. Note that the unimodular matrix is defined as follows.

Definition 11.1 *A nonsingular integer matrix M is unimodular if* $\det(M) = \pm 1$.

It then follows that a nonsingular integer matrix Z is unimodular if and only if Z^{-1} is an integer matrix. The original LLL algorithm computes $\widehat{R}$ in the form of DU, where D is a positive diagonal scaling matrix and U is upper triangular with ones on its diagonal. Moreover, the columns of $\widehat{R} = DU$ form a reduced basis in the following sense.

Definition 11.2 *The columns of an upper triangular matrix $\widehat{R} = DU$ form a reduced basis if*

$$|u_{i,j}| \leq 0.5, \quad \textit{for} \quad 1 \leq i < j \leq n, \tag{11.4}$$

and

$$d_i^2 \geq (\omega - u_{i-1,i}^2)d_{i-1}^2, \quad \textit{for} \quad 2 \leq i \leq n, \tag{11.5}$$

where $0.25 < \omega < 1$ is a parameter that controls the rate of the convergence of the LLL algorithm.

Condition (11.4) states that the absolute value of all off diagonal elements of U are relatively small. Condition (11.5) states that the diagonal elements of U are ordered in a certain manner.

Substituting R in (11.2) with (11.3), we get

$$\min_{\mathbf{x}\in\mathbb{Z}^n} \|Q\widehat{R}Z^{-1}\mathbf{x} - \widehat{\mathbf{b}}\|_2^2.$$

Since Q is orthogonal and Z is integer unimodular, the above problem is equivalent to

$$\min_{\mathbf{y}\in\mathbb{Z}^n} \|\widehat{R}\mathbf{y} - \tilde{\mathbf{b}}\|_2^2, \tag{11.6}$$

where $\mathbf{y} = Z^{-1}\mathbf{x}$ and $\tilde{\mathbf{b}} = Q^T\widehat{\mathbf{b}}$. We call (11.6) *the reduced triangular integer least squares problem* since the columns of $\widehat{R}$ form a reduced basis.

The orthogonal matrix Q in (11.3) is not explicitly computed in the original LLL algorithm. In some applications, Q is not required. However, in the problem of integer least squares, Q is required, as shown above. Moreover, the orthogonality of Q is crucial for the accuracy of the solution. Luk and Tracy [4] presented an improved LLL algorithm based on plane reflections. Their algorithm explicitly computes the matrix Q. In this chapter, we compare the orthogonality of Q produced by the original LLL algorithm and the improved LLL algorithm and show that the improved algorithm produces far superior orthogonal Q over that produced by the original algorithm. The chapter is organized as follows. The original and the improved LLL algorithm are described in section 11.2 and section 11.3 respectively. The orthogonality of the matrix Q produced by both algorithms are compared in section 11.4. We demonstrate our numerical experiments in section 11.5.

11.2 Original LLL algorithm

How does the LLL algorithm compute the QRZ decomposition (11.3) of a nonsingular and upper triangular matrix R? Recall that in the original LLL algorithm R is given in the form of DU, where D is positive diagonal and U has unit diagonal.

To impose the condition (11.4) on $U \equiv [u_{i,j}]$, we begin by defining the elementary unimodular transformation. Let $\mathbf{e}_i, \mathbf{e}_j \in \mathbb{Z}^n$, $i < j$, denote the ith and jth unit coordinate vectors respectively, then we define the elementary unimodular matrix

$$M_{ij} \equiv I - \gamma\mathbf{e}_i\mathbf{e}_j^{\mathrm{T}}, \tag{11.7}$$

where γ is an integer and I is the identity matrix. It can be verified that M_{ij} is an integer unimodular matrix. Suppose that $|u_{i,j}| > 0.5$ for some i and j. Set γ in (11.7) to be the integer closest to $u_{i,j}$:

$$\gamma = \lceil u_{i,j} \rfloor \tag{11.8}$$

and apply M_{ij} to U:

$$U \longleftarrow UM_{ij}.$$

It can be verified that the (i,j)th element of the new U satisfies the condition (11.4). We define the following procedure.

Procedure 11.1 (`Decrease`(i,j)) *Given U, calculate γ by (11.8) and set M_{ij} by (11.7). Then apply M_{ij} to U and accumulate M_{ij} in a matrix Z initialized to the identity:*

$$U \longleftarrow UM_{ij}, \quad Z \longleftarrow ZM_{ij}.$$

Now we consider the condition (11.5). Suppose that (11.5) is not satisfied for some i:

$$d_i^2 < (\omega - u_{i-1,i}^2)d_{i-1}^2, \quad 2 \le i \le n.$$

Denoting Π_i as the permutation in the $(i-1, i)$ plan, we interchange the columns $i-1$ and i of U:

$$U \longleftarrow U\Pi_i.$$

To restore the upper triangular structure of the permuted U, we define

$$X_i \equiv \begin{bmatrix} I_{i-2} & & & \\ & \mu & 1-\xi\mu & \\ & 1 & -\xi & \\ & & & I_{n-i} \end{bmatrix}, \tag{11.9}$$

where ξ and μ are parameters to be determined to restore the upper triangular structure in the new U and I_k denotes the k-by-k identity matrix. It can be verified that

$$\det(X_i) = -1$$

and

$$X_i^{-1} = \begin{bmatrix} I_{i-2} & & & \\ & \xi & 1-\xi\mu & \\ & 1 & -\mu & \\ & & & I_{n-i} \end{bmatrix}. \tag{11.10}$$

The matrix X_i^{-1} is made up of a product of two Gauss transformations. Here is a quick illustration:

$$\begin{bmatrix} \xi & 1-\xi\mu \\ 1 & -\mu \end{bmatrix} = \begin{bmatrix} 1 & \xi \\ 0 & 1 \end{bmatrix} \begin{bmatrix} 0 & 1 \\ 1 & -\xi \end{bmatrix}.$$

This matrix X_i^{-1} is a workhorse in the LLL algorithm and the following relation is the key:

$$\begin{bmatrix} \xi & 1-\xi\mu \\ 1 & -\mu \end{bmatrix} \begin{bmatrix} 1 & \mu \\ 0 & 1 \end{bmatrix} \begin{bmatrix} 0 & 1 \\ 1 & 0 \end{bmatrix} = \begin{bmatrix} 1 & \xi \\ 0 & 1 \end{bmatrix}. \tag{11.11}$$

The above equation says that the matrix X_i^{-1} restores the triangularity of a permuted 2-by-2 triangular matrix. Note that both upper triangular matrices in (11.11) have ones on their diagonals.

The paper [2] gives the values of ξ and μ in (11.10). As is obvious from (11.11), μ is given by

$$\mu = u_{i-1,i}. \tag{11.12}$$

In addition, ξ is given by

$$\xi = \frac{\mu d_{i-1}^2}{d_i^2 + \mu^2 d_{i-1}^2}. \tag{11.13}$$

Lenstra et al. [2] give the formulas that are used to update the squares of the diagonal elements d_{i-1} and d_i of D. Specifically,

$$\hat{d}_{i-1}^2 = d_i^2 + \mu^2 d_{i-1}^2 \quad \text{and} \quad \hat{d}_i^2 = (d_i^2 d_{i-1}^2)/\hat{d}_{i-1}^2, \tag{11.14}$$

where $\hat{d}_{i-1}$ and $\hat{d}_i$ are the new diagonal elements.

The actions to enforce the condition (11.5) are written out in the next procedure.

Procedure 11.2 (`SwapRestore`(i)) *Given D^2, U, and Z, update D^2, swap columns $i-1$ and i of U, apply the transformation X_i^{-1} to the permuted U to restore the upper triangular structure, and update Z:*

$$D^2 \longleftarrow \widehat{D}^2, \quad U \longleftarrow X_i^{-1}U\Pi_i, \quad \text{and} \quad Z \longleftarrow Z\Pi_i.$$

The matrix $\widehat{D}^2$ is obtained by (11.14) *and X_i^{-1} is computed by the equations* (11.10), (11.12), *and* (11.13).

Now, we present the original LLL algorithm.

Algorithm 11.1 (LLL) *Given D^2, U, this algorithm computes new D, U, and Z in the QRZ decomposition* (11.3) *so that the columns of the new U form a reduced basis.*

```
Z ← I;
k ← 2;
while k ≠ n
    if |u_{k-1,k}| > 0.5
        Decrease(k − 1, k);
    endif
    if d_k^2 < (ω − u_{k-1,k}^2) d_{k-1}^2
        SwapRestore(k);
        k ← max(k − 1, 2);
    else
        for i = k − 2 down to 1
            if |u_{i,k}| > 0.5
                Decrease(i, k);
            endif
            k ← k + 1;
        endfor
    endif
endwhile
```

It is shown in [1] that the LLL algorithm is effective in reducing the condition number of most ill-conditioned matrices that occur in practice.

11.3 Improved LLL algorithm

In [4], Luk and Tracy presented an improved LLL algorithm. Their algorithm computes the QRZ decomposition (11.3) of R instead of the product form DU as in the original LLL algorithm. Thus they redefined the reduced basis in terms of R instead of U in Definition 11.2.

Definition 11.3 *The columns of an upper triangular matrix R form a reduced basis if*

$$r_{i,i} \geq 2\,|r_{i,j}|, \quad \text{for} \quad 1 \leq i < j \leq n, \tag{11.15}$$

and

$$r_{i,i}^2 \geq \left[\omega - (r_{i-1,i}/r_{i-1,i-1})^2\right] r_{i-1,i-1}^2, \qquad \text{for} \qquad 2 \leq i \leq n, \tag{11.16}$$

where $0.25 < \omega < 1$ is a parameter that controls the rate of convergence.

As we can see, the above definition of reduced basis is a scaled version of Definition 11.2. Suppose that (11.15) is not satisfied for some i and j, i.e.,

$$r_{i,i} < 2|r_{i,j}|.$$

Setting

$$\gamma = \lceil r_{i,j}/r_{i,i} \rfloor, \tag{11.17}$$

We construct the unimodular matrix M_{ij} by (11.7), and then apply M_{ij} to R:

$$R \longleftarrow RM_{ij}.$$

At the same time, we accumulate the transformations M_{ij} in Z:

$$Z \longleftarrow ZM_{ij}.$$

It can be verified that the (i,j)th element of the new R satisfies the condition (11.15).

Procedure 11.3 (`NewDecrease`(i,j)) *Given R, calculate γ by* (11.17) *and set M_{ij} by* (11.7). *Then apply M_{ij} to R and accumulate M_{ij} in a matrix Z initialized to the identity:*

$$U \longleftarrow RM_{ij}, \quad Z \longleftarrow ZM_{ij}.$$

For the condition (11.16), their algorithm uses the plane reflection in the $(i-1,i)$-plane, $2 \le i \le n$, of the form:

$$J_i \equiv \begin{bmatrix} I_{i-2} & & & \\ & c & s & \\ & s & -c & \\ & & & I_{n-i} \end{bmatrix} \quad \text{where} \quad c^2 + s^2 = 1. \tag{11.18}$$

Note that, just as $\det(X_i) = -1$ in the original LLL algorithm, we have

$$\det(J_i) = -1.$$

However, the matrix J_i is orthogonal. The implication of using orthogonal J_i instead of nonsingular X_i in (11.9) will be discussed in section 11.4. Suppose that the condition (11.16) is not satisfied for some i, i.e.,

$$r_{i,i}^2 < \left[\omega - (r_{i-1,i}/r_{i-1,i-1})^2\right] r_{i-1,i-1}^2.$$

We interchange the columns i and $i-1$ of R:

$$R \longleftarrow R\Pi_i,$$

and apply a plane reflection J_i in (11.18) to restore the upper triangular structure:

$$R \longleftarrow J_i R.$$

We can then update Z, and accumulate the transformations J_i in Q:

$$Z \longleftarrow Z\Pi_i, \qquad Q \longleftarrow QJ_i.$$

Now we define the following procedure and present the improved LLL algorithm.

Procedure 11.4 (`NewSwapRestore`) *Given R, Z, and Q, swap the columns $i-1$ and i of R and those of Z, apply a plane reflection J_i (11.18) to restored the permuted R back to upper triangular form, and update Z and Q:*

$$R \longleftarrow J_i R \Pi_i, \quad Z \longleftarrow Z\Pi_i, \quad \textit{and} \quad Q \longleftarrow QJ_i.$$

Algorithm 11.2 (Improved LLL) *Given R, this algorithm computes new R, Q, and Z in the QRZ decomposition and the columns of the new R form a reduced basis.*

```
Q ← I; Z ← I;
k ← 2;
while k ≠ n
    if |r_{k-1,k-1}| < 2|r_{k-1,k}|
        NewDecrease(k − 1, k);
    endif
    if r_{k,k}^2 < [ω − (r_{k-1,k}/r_{k-1,k-1})^2] r_{k-1,k-1}^2
        NewSwapRestore(k);
        k ← max(k − 1, 2);
    else
        for i = k − 2 down to 1
            if |r_{i,i}| < 2|r_{i,k}|
                NewDecrease(i, k);
            endif
            k ← k + 1;
        endfor
    endif
endwhile
```

11.4 Comparison

In this section, we compare the two LLL algorithms: the original and the improved one. In particular, we compare the orthogonality of Q computed by both algorithms, since, as shown at the end of section 11.1, the orthogonality of Q is crucial for the accuracy of the computed least squares solution.

The improved LLL algorithm computes Q by accumulating plane reflection as shown in Algorithm 11.2, whereas the original LLL algorithm does not explicitly produce Q. We first present two possible ways of modifying Algorithm 11.1 to produce Q.

First, suppose that Algorithm 11.1 outputs $\widehat{R}$ in the form of $\widehat{D}\widehat{U}$. Then from (11.3), we have

$$Q = RZ\widehat{R}^{-1} = (DU)Z(\widehat{D}\widehat{U})^{-1}.$$

Thus Q can be computed at the end of Algorithm 11.1 by solving a sequence of upper triangular systems. One may argue that this method is obviously undesirable, because the orthogonality of the computed Q is expected to be poor when $\widehat{R} = \widehat{D}\widehat{U}$ is ill-conditioned.

Then we present the second, not so obvious, method for computing Q in Algorithm 11.1 by accumulating the X_i matrices constructed in `SwapRestore` procedure.

Let $D_0 = D$ and $U_0 = U$ initially. Suppose that at the end of the ith iteration, the algorithm has D_i and U_i stored in D and U, respectively, and the program terminates after

N iterations. Thus, $D_N = \widehat{D}$ and $U_N = \widehat{U}$. Also, suppose $U_i = X_i^{-1} U_{i-1} Z_i$, where X_i is constructed in `SwapRestore` procedure in the ith iteration. If the procedure is not called in the ith iteration, then $X_i = I$. The matrix Z_i is a combination of the matrices M_{ij} in (11.7) and Π_i constructed in the ith iteration. So, when the program terminates, $Z = Z_1 \cdots Z_N$. Thus, we have

$$U_N = (X_1 \cdots X_N)^{-1} U_0 (Z_1 \cdots Z_N) = (X_1 \cdots X_N)^{-1} U_0 Z. \tag{11.19}$$

From (11.3), we have

$$D_0 U_0 = Q D_N U_N Z^{-1}.$$

Substituting U_N in the above equation with (11.19), we get

$$D_0 U_0 = Q D_N (X_1 \cdots X_N)^{-1} U_0,$$

which implies that

$$Q = D_0 (X_1 \cdots X_N) D_N^{-1} = D (X_1 \cdots X_N) \widehat{D}^{-1}.$$

Now, we integrate the computation of Q into Procedure 11.2 and Algorithm 11.1.

Procedure 11.5 (`Modified SwapRestore`(i)) *Given D^2, U, Q, and Z, swap columns $i-1$ and i of U, apply the transformation X_i^{-1} to the permuted U to restore the upper triangular structure, and update Q and Z:*

$$D^2 \longleftarrow \widehat{D}^2, \quad U \longleftarrow X_i^{-1} U \Pi_i, \quad Q \longleftarrow Q X_i, \quad \textit{and} \quad Z \longleftarrow Z \Pi_i.$$

The matrix $\widehat{D}^2$ is obtained by (11.14) *and X_i^{-1} and X_i are computed by the equations* (11.10), (11.9), (11.12), *and* (11.13).

Algorithm 11.3 (Modified LLL) *Given D^2, U, this algorithm computes new D, U, and Z in the QRZ decomposition* (11.3) *so that the columns of the new U form a reduced basis.*

```
Z ← I; Q ← D;
k ← 2;
while k ≠ n
    if |u_{k-1,k}| > 0.5
        Decrease(k − 1, k);
    endif
    if d_k^2 < (ω − u_{k-1,k}^2) d_{k-1}^2
        SwapRestore(k);
        k ← max(k − 1, 2);
    else
        for i = k − 2 down to 1
            if |u_{i,k}| > 0.5
                Decrease(i, k);
            endif
            k ← k + 1;
        endfor
    endif
endwhile
Q ← QD^{-1};
```

11.5 Numerical experiments

In this section, we present our numerical experimental results for comparing the orthogonality of Q produced by the original and the improved LLL algorithms.

We program the algorithms in Matlab and run our experiments on a 2GHz laptop with 2GB RAM, Windows Vista Home Basic SP1 operation system. In the experiments, we generate 10×10 random matrices with condition numbers 10^k, for $k = 0, 1, \ldots, 16$, as follows. First, we generate a diagonal singular value matrix D, whose diagonal elements are linearly spaced between 1 and 10^{-k}. Then we obtain two random orthogonal matrices U and V from the QR decomposition of two random matrices. Finally, we form $A = UDV^T$. Thus $\text{cond}(A) = 10^k$. We measure the orthogonality of Q by the norm $\|Q^TQ - I\|_2$.

For each random matrix A, the matrix Q was computed by the three different ways described in the previous sections. Specifically, in Algorithm 11.2, Q was computed by accumulating the plane reflections. In the original LLL algorithm, Q was computed from $Q = RZ\widehat{R}^{-1}$ and, alternatively in Algorithm 11.3, from the accumulation of X_i^{-1} as described in the previous section.

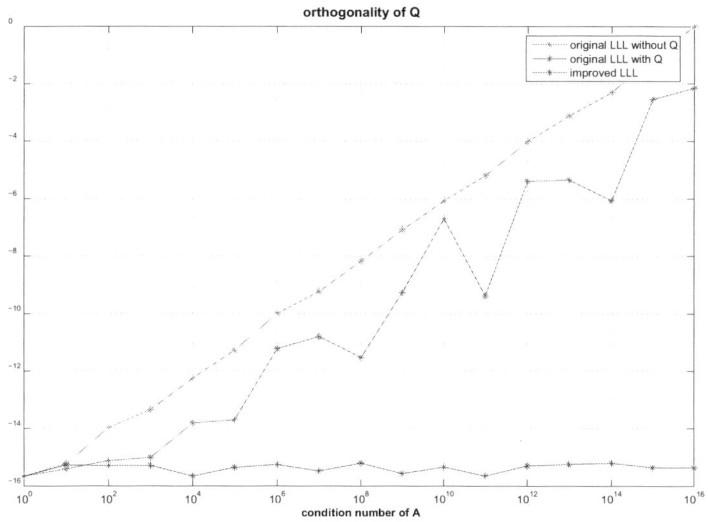

The figure above shows a typical result of our experiments. We can see that the orthogonality of Q computed by the original LLL algorithm is approximately proportional to the condition number of A, whereas the orthogonality of Q computed by the improved LLL algorithm remains good independent of the condition number of A.

11.6 Conclusion

In this chapter, we consider the original LLL algorithm and the improved LLL algorithm by comparing especially the orthogonality of Q computed by both methods. We find that the improved LLL algorithm produces much better results than that of the original LLL algorithm from the viewpoint of orthogonality, which is desirable in applications such as solving integer least squares problems.

References for Chapter 11

[1] B. Hassibi and H. Vikalo, *On the sphere-decoding algorithm I: Expected complexity*, IEEE Tran. Signal Processing, 53 (2005), pp. 2806–2818.

[2] A. K. Lenstra, H. W. Lenstra, and L. Lovasz, *Factoring polynomials with rational coefficients*, Mathematische Annalen, 261 (1982), pp. 515–534.

[3] F. T. Luk and S. Qiao, *Numerical properties of the LLL algorithm*, Advance Signal Processing Algorithms, Architectures, and Implementations XVII, Volume 6697 of Proceeding of SPIE, pages 6697–3, edited by Franklin T. Luk, The International Society of Optical Engineering, 2007.

[4] F. T. Luk and D. M. Tracy, *An improved LLL algorithm*, Linear Algebra and Its Applications, 428 (2008), pp. 441–452.

[5] S. Qiao, *Integer least squares: sphere decoding and the LLL algorithm*, Proceedings of C3S2E-08, ACM International Conference Proceedings Series, edited by Bipin C. Desai, Concordia University, Montreal, QC, May 12–13, 2008, pp. 23–28.

Made in United States
Orlando, FL
04 November 2026

90201166R00077